Career Opportunities in Clinical Drug Research

ALSO FROM COLD SPRING HARBOR LABORATORY PRESS

At the Bench: A Laboratory Navigator, Updated Edition

At the Helm: Leading Your Laboratory, 2nd Edition

Career Opportunities in Biotechnology and Drug Development

Experimental Design for Biologists

Lab Dynamics: Management Skills for Scientists

Lab Math: A Handbook of Measurements, Calculations, and Other Quantitative Skills for Use at the Bench

Lab Ref, Volume 1: *A Handbook of Recipes, Reagents, and Other Reference Tools for Use at the Bench*

Lab Ref, Volume 2: *A Handbook of Recipes, Reagents, and Other Reference Tools for Use at the Bench*

A Short Guide to the Human Genome

Statistics at the Bench: A Step-by-Step Handbook for Biologists

Career Opportunities in Clinical Drug Research

Rebecca J. Anderson

COLD SPRING HARBOR LABORATORY PRESS
Cold Spring Harbor, New York

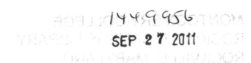

Career Opportunities in Clinical Drug Research

©2010 by Cold Spring Harbor Laboratory Press, Cold Spring Harbor, New York
Printed in the United States of America

Publisher and Acquisition Editor	John Inglis
Director of Development, Marketing, & Sales	Jan Argentine
Developmental Editor	Judy Cuddihy
Project Manager	Mary Cozza
Production Editor	Mala Mazzullo
Production Manager	Denise Weiss
Book Marketing Manager	Ingrid Benirschke
Sales Account Managers	Jane Carter and Elizabeth Powers
Cover Designer	Armen Kojoyian

Library of Congress Cataloging-in-Publication Data

Anderson, Rebecca J. (Rebecca Jane), 1949-
    Career opportunities in clinical drug research / by Rebecca J. Anderson.
        p. ; cm.
    Includes index.
    ISBN 978-1-936113-05-7 (hardcover : alk. paper)
    1. Drugs--Testing--Vocational guidance. 2. Clinical pharmacology--Vocational guidance. I. Title.
    [DNLM: 1. Career Choice. 2. Drug Industry. 3. Biotechnology. 4. Clinical Trials as Topic. 5. Research Personnel. 6. Vocational Guidance. QV 736 A549c 2010]

    RM301.27.A53 2010
    615'.19023--dc22
                                        2010017643

10 9 8 7 6 5 4 3 2 1

All World Wide Web addresses are accurate to the best of our knowledge at the time of printing.

All Cold Spring Harbor Laboratory Press publications may be ordered directly from Cold Spring Harbor Laboratory Press, 500 Sunnyside Blvd., Woodbury, New York 11797-2924. Phone: 1-800-843-4388 in Continental U.S. and Canada. All other locations: (516) 422-4100. FAX: (516) 422-4097. E-mail: cshpress@cshl.edu. For a complete catalog of all Cold Spring Harbor Laboratory Press publications, visit our website at http://www.cshlpress.com/.

To the patients who are waiting

Contents

Preface

As hiring managers, my colleagues and I have always struggled to fill open clinical positions, despite an abundance of applications. We sift through mounds of résumés from eager and talented applicants, including many from the research laboratories of our own companies. Unfortunately, we reluctantly reject most of those applicants, because they lack one important prerequisite: clinical experience. Even for entry level positions, for which such experience is not absolutely required, we prefer to hire people who have an understanding of clinical studies. Those who don't are at a significant disadvantage.

Clinical experience is not just another item in the job posting for drug, medical device, and contract research organization (CRO) companies. These companies are obligated by regulatory requirements to employ only clinical workers who are qualified for the jobs they hold and to ensure that they update those qualifications through internal training programs. For new employees who lack clinical experience, the company (and more specifically, the hiring manager) must invest considerable resources in clinical training before the workers can be assigned official responsibilities on clinical studies.

Clearly, physicians, nurses, and pharmacists have acquired relevant experience and easily meet this job prerequisite. However, most of the positions in industrial clinical departments, particularly at the entry levels, do not require a medical degree or healthcare licensure. What is particularly frustrating to us, as hiring managers, is that undergraduate and graduate students can acquire, in a number of ways, clinical experience that we will accept, and it is not difficult to obtain. If only someone would explain to job seekers how important clinical experience is and provide information on how to obtain it.

A second frustration for hiring managers is that many talented and potentially interested candidates are simply unaware of the job opportunities in industrial clinical departments. Academia generally does not inform or prepare students for careers in research and development of new medical products. It is simply amazing to me that so many of my industry coworkers stumbled upon their careers in clinical research, like I did, merely by chance. In most cases, once those opportunities were presented to us, we found fulfilling and successful careers. Undoubtedly, other eager job seekers would also find this career path attractive. If only someone would tell them about it.

For those with an interest in science and a desire to find better treatments for patients, this book will pull back the curtain on rewarding and lucrative career opportunities that are rarely mentioned in academic degree programs. For those who were lucky enough to discover clinical jobs in industry, this book will explain strategies for landing an entry level job as an outsider—including, specifically, how to get that all-important clinical experience.

The book covers clinical research that is conducted at pharmaceutical, biotechnology, medical device, and CRO companies. At various times in my career, I have had the privilege of working in the research and development divisions of all of those industry segments. However, I am indebted to a number of coworkers and professional colleagues who assisted me in various capacities and without whom this book would not have been possible.

First and foremost, I am deeply grateful to Susan Aiello and Diana Patten, who freely offered their insight, guidance, and encouragement throughout the entire course of this project, from development of the original concept to completion of the book. I am equally grateful to Julie Silver and the faculty of her non-fiction book course at Harvard's Department of Continuing Education. The course provided invaluable advice on all aspects of non-fiction writing and book production, and it inspired me to push my writing ambitions to levels I had not previously considered.

The following individuals graciously provided input and perspective on various topics covered in the book and verified specific factual information: Kathleen Block, Laura Bloss, Patti Brennan, David Mee, Angela Meisterling, Betty Mendes, Diana Quinto, Peggy Smith, John Tiso, David Townson, and Kristina Welch.

I am also grateful to the following subject area experts who carefully critiqued the chapters related to their professional expertise, provided many helpful suggestions, and verified the content: Frances Akelewicz, Shelia M. Brown-Walker, John Constant, Ann Dugan, Marta Fields,

MaryAnn Foote, Nicole Harasym, Sue Hudson, Donna Jacobs, Susan Lyman, Carol Marimpietri, Patricia Mighetto, Mary Bernadette Ott, Heidi Reidies, Rob Tarney, and James Yuen.

Finally, I owe a special thanks to my Developmental Editor, Judy Cuddihy, who has patiently guided me through the editorial process, offered many helpful suggestions for improving the book's content, and explained the finer points of book publishing. Due to her efforts, the book is much better than it otherwise would have been. I also thank the staff at Cold Spring Harbor Laboratory Press, including Publisher John Inglis, Project Manager Mary Cozza, Production Editor Mala Mazzullo, Director of Development, Marketing, and Sales Jan Argentine, and Production Manager Denise Weiss for their support and exceptional efforts in producing the book.

Abbreviations and Acronyms

AAMI	Association for the Advancement of Medical Instrumentation
ACRO	Association of Clinical Research Organizations
ACRP	Association of Clinical Research Professionals
AE	adverse event
AMWA	American Medical Writers Association
APhA	American Pharmacists Association
ASA	American Statistical Association
ASQ	American Society of Quality
BIO	Biotechnology Industry Organization
CDISC	Clinical Data Interchange Standards Consortium
CDM	clinical data management
CEU	Continuing Education Unit
CFR	Code of Federal Regulations
CHPA	Consumer Healthcare Products Association
CMC	chemistry, manufacturing, and control
CME	Continuing Medical Education
CNU	Continuing Nursing Unit
CQA	clinical quality assurance
CRA	clinical research associate
CRF	case report form
CRO	contract research organization
CSR	clinical study report
CSS	Clinical Summary of Safety
CTA	Clinical Trial Authorization
CTD	Common Technical Document
DIA	Drug Information Association

DMC	data monitoring committee
DMP	data management plan
EC	ethics committee
eCTD	Electronic Common Technical Document
EDC	electronic data capture
EMEA	European Medicines Agency
EU	European Union
FDA	Food and Drug Administration
FDLI	Food and Drug Law Institute
GCP	Good Clinical Practice
GLP	Good Laboratory Practice
GMP	Good Manufacturing Practice
HR	human resources
IB	investigator's brochure
IBS	International Biometric Society
ICF	informed consent form
ICH	International Conference on Harmonization
IDE	Investigational Device Exemption
IEC	independent ethics committee
IND	Investigational New Drug Application
IRB	institutional review board
ISO	International Organization for Standardization
ISMPP	International Society of Medical Publication Professionals
ISPE	International Society for Pharmacoepidemiology
IVRS	interactive voice response system
MAA	Marketing Authorization Application
MD&DI	medical device and diagnostic industry
MD&M	medical design and manufacturing
MedDRA	Medical Dictionary for Regulatory Activities
NCR	no carbon required
NDA	New Drug Application
PhRMA	Pharmaceutical Research and Manufacturers of America
PI	principal investigator
PMA	Premarket Approval

PMI	Project Management Institute
PSI	Statisticians in the Pharmaceutical Industry
QA	quality assurance
QC	quality control
R&D	research and development
RAC	Regulatory Affairs Certification
RAPS	Regulatory Affairs Professional Society
SAE	serious adverse event
SAP	statistical analysis plan
SCDM	Society for Clinical Data Management
SCE	Summary of Clinical Efficacy
SCS	Summary of Clinical Safety
SOP	standard operating procedures
SoCRA	Society of Clinical Research Associates
SQA	Society of Quality Assurance
SQL	Structure Query Language
STC	Society for Technical Communication
TQM	Total Quality Management
UADE	unanticipated adverse device effect
UAT	User Acceptance Testing

1

What Is the Clinical Environment in Industry?

WANT A JOB THAT PAYS WELL, has excellent benefits, and offers fast-track promotions? Want to work for an employer who provides a pleasant work environment with state-of-the-art equipment, encourages on-the-job training, and pays your travel expenses? And on top of that, how about a job aimed at improving patients' health and sometimes saving their lives? If you want to go home every night with a warm feeling inside—and get paid for it—then this book is for you. Working on clinical studies gives you all these perks and offers you a satisfying and rewarding career.

Before any new drug can be marketed, it must undergo extensive clinical testing (i.e., studies using patients) to demonstrate that it is safe and that it has therapeutic value. The makers of medical devices must also conduct studies in patients to evaluate the safety and performance of products such as pacemakers, heart valves, and artificial hips.

Lots of people work together to conduct a clinical study, and most of them are not physicians. You can qualify for many jobs in clinical fields, such as clinical operations, data management, quality assurance, and regulatory affairs, with just a bachelor's or graduate degree. But there's a catch. To qualify for most clinical research jobs in the medical products industry, you need to have prior clinical experience; hiring managers won't even consider you without it. So, how do you get started?

This book will show you how to qualify for those jobs and launch a lucrative, rewarding clinical career developing new drugs or medical devices. Part I explains how drug and medical device companies conduct clinical studies, with an emphasis on the jobs that are easiest for you to enter as an outsider. In Part II, you will learn what people in each of those entry-level positions do, what qualifications you need, and most importantly how to get "clinically relevant" experience that a hiring manager

1

will accept. Part III gives you practical advice for conducting a successful job search and landing your first clinical job.

As you will quickly see, the world of clinical studies has its own terminology. Each chapter explains technical terms as they are introduced. In addition, a glossary is included for your reference. (Building your vocabulary of clinical study terms will give you another advantage during your job search. After all, hiring managers expect you to have a basic understanding of the terms they use every day.)

WHERE ARE THE JOBS?

The companies that are required by regulatory agencies to conduct clinical studies fall into three categories: pharmaceutical, biotechnology, and medical device companies. Pharmaceutical companies (sometimes called big pharma companies) are mostly large, multinational companies. The drugs that they manufacture and sell are organic chemicals, which they either synthesize by chemical reactions in the laboratory or extract from natural sources such as plants. Biotechnology companies (or biotechs) make and sell drugs that are manufactured using molecular biology and genetic engineering methods. Scientists create biotech drugs by modifying biological substances (such as proteins or antibodies) from their natural form and manufacture the product using biological fermentation methods.

Many companies now develop both biotech and chemically synthesized drugs, thus blurring the distinction between pharmaceuticals and biotechnology. Biopharmaceutical (or biopharma) is a collective term that refers to both biotechnology and pharmaceutical companies and the drugs they produce. Most biopharmaceutical companies have their headquarters or a major branch office in the United States, which is where most of the opportunities for clinical development jobs are located.

Medical device companies make and sell nondrug products for patient use, ranging from home pregnancy tests to treadmills to heart-lung machines. Of these, Class I and II devices are those whose failure or misuse have little or no impact on patient care, are unlikely to cause direct patient injury, and therefore generally are not tested in clinical studies. On the other hand, Class III medical devices, whose failure or misuse is reasonably likely to cause serious patient injury (such as fetal monitors, silicone breast implants, and asthma inhalers), require clinical testing similar to that conducted by drug companies. Most medical device companies are small, entrepreneurial companies and develop one, or only a few, products. A few, large medical device companies develop and sell a wide range of products for hospital or home use.

Biopharmaceutical and medical device companies sometimes need additional workers to carry out their scheduled clinical studies. Rather than hire additional staff to meet short-term or specialized needs, they typically turn to contract research organizations (CROs). As the name implies, CROs lend the services of their trained employees to client companies; they do not develop their own medical products. Because the medical products industry is highly regulated, CRO employees must have the same qualifications and be able to conduct clinical studies with the same standards as their clients. CROs range in size from small, specialty contractors to large, full-service companies. Most of the large, multinational CROs have their headquarters or a major affiliated office in the United States, which is where most of the opportunities for clinical jobs are located.

Biopharmaceutical, medical device, and CRO companies always need workers to conduct clinical studies. Once a study starts, a drug or medical device company is committed to follow through until the study's completion, and many clinical studies continue for more than a year. When biopharmaceutical and medical device companies are thriving and growing, they expand their clinical departments to support development of their new products. When those companies are struggling financially, they will cut back their budgets and perhaps lay off staff. However, they must continue to support their ongoing and planned clinical studies. When internal resources are limited, they engage CROs to conduct or continue the studies on their behalf. CROs thrive and grow to accommodate increased requests for their services, sometimes with remarkable growth of their workforce.

Thus, despite wide fluctuations in the economy and other factors, openings for clinical jobs in medical product development are always available. Sometimes, most jobs will be at biopharmaceutical and medical device companies; at other times, the jobs will be at CROs. But there will be jobs.

INDUSTRY CLIMATE

The Research and Development (R&D) division of biopharmaceutical and medical device companies encompasses three separate work cultures.

Each is driven by the nature of the work, which differs in its goals and the methods used to achieve them.

Research scientists and engineers who work in industry laboratories are charged with finding novel and innovative products to treat or diagnose disease. Their research activities are not very different from academic or independent research laboratories, except that they are expected to work toward the goal of discovering a therapeutically useful product. They are encouraged to be creative, explore the latest frontiers of science, and take risks. Except for rules that protect the environment and their own safety, their work is not constrained by regulatory requirements or company-imposed standard procedures.

A second group of laboratory scientists and engineers is charged with taking a discovery made by the research scientists and assessing its potential as a commercial product. This preclinical assessment is the first stage of a new product's development. These development scientists and engineers collect a well-defined set of data that describe the product's characteristics and performance in a controlled laboratory setting, including tests in animals. The work environment is much more structured than research laboratories, and workers follow regulatory requirements for Good Laboratory Practices (GLPs), which not only stipulate the required design features of the experiments but also the rules for recordkeeping, equipment calibrations, and controlling experimental variables.

The third segment, and by far the largest, longest, and most labor-intensive aspect of R&D, is the work of the clinical department. Healthcare professionals and many other skilled people work in teams to take products that survive preclinical assessment and determine their therapeutic value in patients. Like the preclinical development environment, the clinical culture is highly structured, and workers follow regulatory requirements for Good Clinical Practices (GCPs), which ensure the safety and welfare of patients during treatment with experimental products, as well as the integrity of the data.

Admittedly, working in an industrial clinical research setting is not for everyone. But for those who are comfortable in this environment, clinical work is personally and professionally rewarding and has many attractive features compared with academic and independent research laboratories. You can expect a collaborative work environment, a supportive but demanding senior management structure, lots of up-side career potential, and great fringe benefits.

BOX 1-1. *Summary of Industry Clinical Climate*

- Good salary, bonuses, and benefits
- Generous study budgets
- More job security for clinical jobs than nonclinical area of R&D
- Coworkers are bright, hardworking, considerate, and talented
- First to see patients benefit from a new treatment
- Have a stake in new treatments that benefits millions of patients
- Resources available to do the job right
- Fast paced work
- Tight deadlines
- Long hours
- Innovation is encouraged but must apply to product development
- Able to present data in top-tier medical journals and at international conferences
- Scientific publication dates are coordinated with business needs and timetables
- Career advancement is based on performance
- Bureaucratic hierarchy
- Decisions are business-based, as well as science-based
- Highly structured and regulated environment
- Flexible, broad-based knowledge is more highly valued than narrow, deep expertise
- Frequent meetings and communication-intensive
- Direct contact with patients is limited
- Reorganizations are common

Collaborative Environment

Clinical studies are a lot like ice hockey. They require a team effort, each person playing a defined position and supporting his or her teammates, to reach the goal. The clock is always ticking. The team must overcome many barriers, sometimes by brute force, to make forward progress. Team members are often sidelined or substituted at critical points along the way. Success is not assured, even when the team does its best. When the clock runs out, the team heads for the showers—perhaps battered and bruised—either to pop the champagne corks or simply to think about the next game.

Yes, it's a rough-and-tumble process. Managing an experimental product through a battery of clinical studies is long and complicated. The work is fast-paced, the competition between companies is keen, the deadlines are tight, and the pressure is constant.

Nevertheless, clinical departments operate in a friendly, informal atmosphere and are generally very pleasant places to work. Clinical investigation of new therapeutic modalities creates a culture founded on highly intellectual research and scientific innovation—the application of cutting-edge science to address human health needs. Everyone is bright, hardworking, considerate, and talented. Employees earn the respect and cooperation of their coworkers, subordinates, and supervisors through their knowledge and helpfulness, not by touting their academic credentials. Clinical teams are motivated to work long hours by the satisfaction of doing something that is truly worthwhile—improving healthcare with new and better treatments.

Everyone in a clinical department shares in a real sense of accomplishment, no matter how small their individual contributions might be. They are the first to see the impact of a new treatment on a patient's medical condition, which may have been previously untreatable. They may see their work presented in medical journals and at world-class medical conferences. Better still, they may rightfully claim a direct contribution to the successful launch of a new commercial product that improves the health of millions of people.

Clinical teams are given a wealth of resources to "do the job right." To supplement the knowledge of their talented colleagues, they regularly invite key opinion leaders from academia and private research institutes to serve on advisory boards or as consultants. Cutting-edge science means that there is no existing blueprint for designing appropriate clinical studies. Clinical teams, therefore, seek advice from the best and brightest medical experts, some of whom may also participate in the studies as clinical investigators. These interactions with external consultants are mutually beneficial: Clinical workers develop a personal rapport with highly regarded medical experts and those experts may gain early access to new treatment modalities for their patients.

In addition, clinical workers can draw on an extensive internal infrastructure to assist them. Librarians and information specialists routinely monitor and alert clinical workers to newly published research findings and the status of competitors' products. Other information technology specialists install and maintain computer software and hardware. Contract and patent attorneys handle the legal aspects of negotiating agreements for external services and protecting the company's intellectual property. Finally, R&D divisions prepare an annual clinical budget to cover the cost of planned clinical studies, including the costs of clerical staff, facilities, and equipment. Finance and accounting experts track clinical study expenses and prepare financial reports. Aside from the annual budget discussions to share their clinical study plans for the coming year, clinical employees usually do not worry about finding funds for their work, unlike their counterparts in academia.

Each clinical study is different, presents its own challenges, and never fails to produce a few surprises no matter how carefully the team has planned it. To keep the product's development on track, clinical teams face a wide range of problems, large and small, some stimulating, and some frustrating. Certainly, there's never a dull moment and the job is never boring.

Developing a new drug or medical device also has a high failure rate. Sometimes, the product fails because of an inherent, insurmountable flaw, even though every team member did everything right. Clinical workers must cherish their successful days amid months of disappointment and distinguish between their personal performance and the product's (sometimes flawed) characteristics. Accepting many disappointments is the price all scientists are willing to pay to savor those few, experimental breakthroughs. Clinical workers in industry must be willing to move on to other projects for business as well as scientific reasons.

Clinical workers are encouraged to be innovative, but their creativity must be channeled along narrow paths. Innovative processes or procedures that save time and money are always valued and rewarded. Sometimes, data collection to answer exploratory questions can be incorporated into the design of essential clinical studies without detracting from the product's development timetable. Such data sometimes can serve as the starting point for a new clinical program, and teams are encouraged to pursue those leads. However, the opportunities to explore open-ended scientific questions are limited.

Clinicians are encouraged to be attentive to unexpected results from their planned studies and to explore the implications of those results. Some of the most successful drugs, such as Viagra®, have come from unanticipated results of clinical studies that were pursuing a different therapeutic use. However, such exploratory work must be conducted with a commercial opportunity in mind, not simply for its value as an intellectual exercise.

Some clinical studies, frankly, are not innovative and only serve to satisfy a regulatory requirement or business need. Examples include comparing an existing drug formulation with a new, improved formulation; determining the proper dose level for various patient groups (such as children vs. the elderly); and testing a product's ease of use (such as two models of blood glucose monitoring kits). However, even in these cases, attentive and bright investigators may uncover a new medical condition (such as patients with a genetic anomaly) or develop innovative computer models to interpret their results.

Decision making is structured in clinical departments. Individuals and teams are empowered to determine the course of their daily work, but senior managers make decisions that have a significant financial impact. Decisions to choose between several alternative experimental lines may fall either to the clinical workers, the senior managers, or both. Although individual input on scientific questions is solicited from all of the internal experts who have a stake in the topic, decisions are made for the good of the company and the best path for developing the product, rather than an individual's preference for pursuing a personally interesting sidelight. The decision making process is not always democratic, and the company's senior managers have the final word.

Clinical departments organize their work in a hierarchy of cross-functional teams. Clinical study teams carry out individual clinical studies. Strategy or project teams coordinate groups of clinical studies needed to develop a single product. Senior management teams oversee, coordinate, and control development of all the products in the company's portfolio.

To oversee and manage the work at all levels and between levels requires extensive, ongoing communication. Large companies are more bureaucratic than small companies, but in all cases, clinical departments have a culture that generates many meetings and incorporates many different viewpoints before making a decision and taking action. Things are less spontaneous than in academia or independent research companies, and clinical workers spend a lot of time in meetings and generate a lot of e-mails, memos, and reports.

However, the structure, hierarchy, and management oversight also provide more career opportunities and options than those available in academia and independent research laboratories. Pay raises, bonuses, and career advancement are based on performance, and performance is judged not so much by what employees achieve individually, but rather by how much they contribute to the success of the teams and projects on which they serve. Industrial employers, therefore, are conscientious about conducting annual performance reviews.

Performance reviews not only document past performance; they also serve as the basis for discussions regarding the employee's future work, and supervisors are expected to assist the people who report to them with their career development. Based on these discussions, workers are encouraged to set job-related and career-enrichment goals for the coming year and plan supplemental training or other activities to strengthen their skills. Through individual coaching, which may be ongoing, supervisors also help employees qualify for promotion or their next career goal.

With regard to career advancement, clinical workers have an advantage over those who work in other areas of R&D. In medical products companies, clinical studies are the most visible aspect of R&D and garner significant attention from influential people in the company. Those who support clinical studies also enjoy greater visibility, and if they make significant contributions, their efforts are more likely to be rewarded, including rapid promotions and other opportunities for high-profile advancement.

Clinical departments put great effort into grooming leaders and managers: People who can inspire others to channel their energy along productive lines. Most companies provide training programs to help clinical workers develop management skills. Those who are willing to take management responsibilities in conjunction with applying their technical expertise are most likely to move to highly responsible and influential positions in industrial clinical departments.

These promotions and high-profile assignments usually move employees further away from day-to-day involvement with clinical studies. Senior managers still assess clinical study results, but they have a broad, strategic vision and responsibility. Rather than concentrating on just one product, study, or clinical skill, they influence the direction of entire clinical programs through their decision making authority.

It's a Business

Companies that develop drugs and medical devices are heavily regulated to protect patients from unnecessary harm and to guide healthcare professionals to use those products correctly. The rules and regulations in clinical departments are extensive and set the boundaries for all of the clinical studies conducted during development of a new medical product. Consequently, there is a disciplined, businesslike aspect to the culture of clinical departments, dictated by the need to plan, authorize, and document each clinical activity.

Ironically, regulatory agencies permit almost no direct contact with patients while a company is developing a new medical product. This ensures patient confidentiality but it prevents company clinical workers from directly observing the product's effects during the study. They must rely on the data they receive from clinical investigators who treat, observe, and assess the study patients at remote clinical sites. (See Chapter 3 for details on investigator and clinical site activities.)

The absence of patient contact allows clinical workers to conduct their businesslike affairs in a relatively comfortable, casual setting. They interact with each other on a first-name basis, regardless of rank. The dress code is casual when they are working in the office or attending internal meetings. However, most companies expect standard business dress when clinical employees represent the company at external meetings or host visitors.

Workers in clinical departments generate a large volume of regulatory documentation. They must follow regulatory rules for collecting specific types of information, checking the content for accuracy, and archiving the documents for future reference. In addition, the company's patent attorneys require documentation to support patent claims and protect the company's intellectual property.

Because so much of the clinical work is conducted to satisfy regulatory requirements, those requirements largely determine the work that individuals and clinical teams must achieve. However, the company's senior managers usually set the deadlines for completing that work based on business needs, and the deadlines are always tight. Furthermore, senior managers may quickly change program priorities or work assignments based on business needs, unexpected study results, or both.

In well-run companies, the senior management's decisions are both business-based and science-based. Senior managers constantly assess which of their products will be the most profitable. If several development programs are competing for resources, managers will favor clinical studies and new product candidates that not only satisfy an unmet medical need but also enhance stockholder value.

The same logic guides their decisions about publically disclosing data from clinical studies. They encourage company scientists and clinicians to publish papers in prestigious medical journals and present their work at high-profile meetings, because public recognition fosters the professional reputations of their employees, as well as the company's public image. However, they often impose constraints on the timing of publication. Release of the data may be authorized only after a development program or clinical study has been discontinued. Positive results may be delayed to coincide with a product launch or patent approval, or rushed to coincide with a stockholder meeting.

At CROs, the clients set the CRO workers' deadlines, which are always shorter than the deadlines imposed within the client company. The client company builds in time to review and, if necessary, ask for corrections of the CRO's work before its own deadline for completing the work. The client company also sets the requirements and scope of the CRO's contract services. Those CRO services may help the client company formulate its business strategy for product development, but the client's strategy is always determined internally. The CRO's senior managers, in turn, set work priorities to keep their most important clients satisfied. CRO managers may quickly change priorities or work assignments to accommodate a demanding client, the unexpected loss of contracted work, or the award of a major, new contract.

In short, the company's financial bottom line always comes first, be it a large pharmaceutical corporation, a small medical device company, or a CRO. The options for creative work schedules such as telecommuting, job sharing, and flextime are more constrained than those allowed by employers in academia or independent research companies. Any consideration for granting a flexible work schedule depends on the nature of the work and whether it serves a business, rather than a personal, need.

Similarly, industry employees can be "hired at will" rather than under a contract, and in this situation their job security is limited. However, because clinical study schedules are less prone to be impacted by changes in the company's balance sheet, clinical positions are generally more secure than those in manufacturing, sales, marketing, and the other divisions of R&D.

The business climate for drug and medical device companies has been volatile in recent years, and the interest in mergers and acquisitions will likely continue. When two or more companies merge, clinical departments are almost always reorganized, and sometimes programs and staff are consolidated.

Decisions to lay off staff as a result of mergers and acquisitions follow a long and well-planned procedure of assessments, communications, and reorganization. In all but the smallest and most volatile companies, employees who are laid off are given a severance package of financial and outplacement assistance, proportional to their years of service.

Even without mergers, acquisitions, or layoffs, clinical departments periodically reorganize in an effort to improve efficiencies in work flow, processes, and procedures. Although job titles, reporting structures, and work units are often changed in these reorganization efforts, the regulatory-driven tasks carried out by clinical workers remain the same and are unaffected.

Whether the company's decision makers impose layoffs or reorganizations, employee flexibility is the key to survival in clinical departments. Workers who have expanded their résumés by acquiring a range of clinical study skills with different types of products in different therapeutic areas are more highly valued than those who have expertise and a deep knowledge of just one thing. Broad-based clinical study knowledge and experience can be leveraged to new opportunities within the company or, if necessary, when applying at a new company.

Salary, Benefits, and Perks

Base salaries at biopharmaceutical, medical device, and CRO companies are generally higher than comparable jobs in academia. Large companies have established pay scales to ensure equity across the organization. Small, entrepreneurial companies have greater flexibility and may offer higher salaries to attract suitably qualified, new employees. Salaries are reviewed and adjusted annually based on the employee's performance, along with cost-of-living adjustments. In companies that meet their corporate goals, employees who perform well can expect annual increases in their base pay of 1%–7%.

In addition to base salaries, whether the pay scales are constrained or not, these companies supplement an employee's income with other types of cash incentives. The guidelines for bonus programs vary widely, but in most companies clinical workers qualify for an annual cash bonus of 5%–20%, depending on their rank and their performance. Privately owned companies are typically more generous in their bonus programs, to compensate for a lack of stock ownership.

In publically traded companies, the bonus structure may be divided between cash awards and stock options. Stock options give employees the opportunity to become a partial owner in the company, which historically has been viewed as another way to encourage employee performance. In small, start-up companies that subsequently do well, the value of stock shares may increase significantly and represent a major portion of the employee's income. However, employees typically must wait several years before they are allowed to exercise their options, during which time the stock value may have fluctuated significantly or be of no net value.

Public companies may also offer their employees the opportunity to purchase stock as part of a savings plan. If the company and stock performance are good, the value of these shares may increase more swiftly than other investment programs. In addition, by purchasing stock, the employee becomes a stockholder with all the privileges of other stockholders.

In addition, most healthcare companies offer generous retirement plans and medical insurance benefits. As an added incentive to participate in these plans, many companies match the employee's contributions to the retirement plan and defray a significant portion of the cost of health insurance. Altogether, the cash equivalent of these employee benefit programs (i.e., bonuses, stock, savings, retirement, and insurance) can equal 40%–60% of the employee's base pay.

As an added incentive to attract new clinical workers, companies often offer additional financial assistance. Because cash bonuses are awarded at the end of a year's service when their performance is assessed, new employees must wait a full year before receiving their first bonus. To compensate for this delay, clinical job candidates may be offered a one-time signing bonus which is awarded to them when they begin their employment. In addition, if the new employee must move to a new location to accept the job, the company often offers relocation assistance. Based on the job level and the urgency in hiring a desirable candidate, the relocation package may cover moving expenses, temporary housing costs, and the real estate costs of selling and purchasing a home.

Drug, medical device, and CRO companies set rules for how much time employees may take off and monitor time off more closely than academic and entrepreneurial employers. Some companies set separate rules for vacation time, sick days, and personal leave; others lump all of these categories together under one "paid time off" heading. New employees are typically awarded a total of 3 weeks of annual paid time off. With increased years of service, additional days of paid leave are usually added to the annual allotment.

Large and mid-size companies also offer their workers a number of other benefits to make the work environment pleasant and encourage employees to do their best. Common fringe benefits include a subsidized on-site cafeteria, generous family leave provisions, and (for employees who wish to continue their education while working) tuition reimbursement. Some companies also provide an on-site fitness center, subsidize daycare, and match the employee's donations to charitable organizations.

Clinical workers at biopharmaceutical, medical device, and CRO companies work in pleasant office space with all the tools needed to do their job. Even at small companies, they benefit from frequently upgraded office equipment, software, and communications devices. In addition, the companies cover the costs and make the arrangements for business-related travel and on-the-job training programs.

Finally, these companies, because they are in the healthcare business, foster "work-life balance" as a core value. However, like most things in life, what you get out of your job depends on what you put into it. The high visibility and importance of clinical work offers clinical workers greater opportunities to be significantly rewarded, but only if they work hard, do their jobs well, and achieve difficult goals. Such workers have certain personal qualities in common.

PERSONAL QUALIFICATIONS FOR CLINICAL POSITIONS

The chapters in Part II of this book discuss the specialized types of work conducted in an industrial clinical department and the qualifications that are especially relevant to that job category. But hiring managers look for a few personal qualifications in all clinical job applicants.

Team Work

Conducting a clinical study or managing a clinical development program requires the coordinated effort of many people, each of whom makes important contributions. Clinical departments therefore organize their work in teams and manage the teams in a matrix that makes clinical workers accountable both to their team leader and their functional unit supervisor.

BOX 1-2. *Summary of Clinical Job Qualifications*

- Work effectively as part of a team
- Take initiative and work interdependently
- Good interpersonal skills
- Effective written and verbal communication skills
- Able to meet deadlines and a sense of urgency
- Able to multitask
- Able to adjust to shifting priorities
- Able to work under defined rules and regulations
- Detail-oriented
- Show good judgment
- Perseverance and persistence
- Satisfied with long-term successes

The work environment is collaborative. If a clinical study or project is to succeed, the team must work cooperatively, sharing information, lending assistance, and solving problems together. Each person not only contributes his or her particular expertise but also coordinates that work with the contributions of other team members.

Whether the team is making plans, carrying out assigned tasks, or solving problems, each team member is expected to actively participate. Discussions may take place in formal team meetings, casual hallway conversations, or electronic messages. Good team members listen actively, ask constructive questions, contribute useful information, and volunteer their services if appropriate. They also share credit, acknowledge the contributions of others, and support the team rather than seeking personal credit at the expense of the team or other team members.

Problem solving is often a shared responsibility, even if the problem comes from one specific technical specialty. The problem and how it is solved may have an impact on the progress of the other team members' efforts. Good clinical workers consider their coworkers' input. Often, the best solution comes from a team member, who is not expert in that field, asking a simple question or making a straightforward observation.

Clinical teams work under constant pressure imposed by time constraints, high expectations, and unforgiving medical conditions.

Inevitably, there are occasional conflicts, but high-performing teams have participants who "work the problem" and avoid placing blame or simply complaining. This requires good interpersonal skills.

Interpersonal Skills

Not everyone who works in a clinical department is an extrovert or "the life of the party." In fact, most clinical workers, given their preference, would work quietly at their desk, avoid conflicts, and take responsibility for solving their own problems. Nevertheless, good clinical team members develop a rapport with their coworkers, value their coworkers' contributions, and earn their coworkers' respect.

In the give-and-take process of solving difficult problems, the team members on high-performing clinical teams focus on the problem, are sensitive to their coworkers' feelings, and engage everyone in the discussion. They realize that their coworkers have a range of personalities and viewpoints, and everyone's perspective is valuable. In addition, good clinical workers sincerely care about others, are good listeners, and offer well-chosen words of support. They foster trust, earn a reputation for being fair, and are easily approachable.

Clinical teams also value coworkers who have a positive, can-do attitude, see obstacles as a challenge, and work hard to overcome them. Valued clinical workers also offer helpful, constructive suggestions and are able to receive criticism gracefully. Many have a pleasant sense of humor.

Clinical teams often operate globally and may include team members representing diverse backgrounds, cultures, and languages. Harmonizing clinical work in such a multinational environment requires not only good interpersonal skills but also knowing how to communicate effectively.

Communication Skills

All clinical workers must be able to express themselves clearly in writing and orally. Their success, both individually and as a team, depends upon relaying accurate and unambiguous information to coworkers and understanding the information they receive from others. Written communications may be short, informal messages sent electronically, or long, formal documents such as regulatory submissions and clinical study reports. A considerable amount of communication in clinical departments takes place orally—discussions in team meetings, telephone conversations, conference calls, and formal presentations to senior management groups.

Whether their communications are formal or informal, clinical workers are expected to exchange information clearly, concisely, and in a timely manner. The most effective workers express their views objectively and defend their point of view without offending others. In discussions, they evaluate and accommodate other points of view to help the clinical team make the best decisions and take the best course of action.

This can be especially challenging when team members are spread across different time zones and cultures, require translations to other languages, and rely on electronic communication tools rather than face-to-face interactions. To understand and be understood under these conditions requires not only accommodating different personality types, a high level of communication skill, and cultural sensitivity, but also a great deal of planning.

The most effective clinical workers are proactive in communicating with others and act on the information they receive. They inform others by stating clearly what they are doing and respond quickly and appropriately to requests from others. Most importantly, they ensure that all stakeholders are informed and engaged. With so many people involved in the complicated process of running clinical studies (e.g., team members, clinical investigators, contractors, vendors, and internal management), clinical workers must be nimble in facilitating the flow of information as they conduct their work.

Multitasking and Flexibility

Although most clinical work is highly structured and must comply with strict regulatory requirements, clinical workers are constantly reacting and adjusting to a very dynamic work environment, and their workday is anything but routine. (Examples of these workday dynamics are included in the chapters that describe specific clinical jobs.) A scientific breakthrough or unexpected clinical results may need to be exploited with urgency. Or, in the course of an ongoing project, a team member may offer a new and better plan. New managers or new market conditions may set new priorities. Projects may end prematurely either because of negative data or a shift in business strategy. All of these situations require workers to be flexible in adjusting their work, maintain their level of enthusiasm, and apply their skills to new situations.

Clinical workers prioritize their work based on its importance and urgency, not solely on their personal level of interest. Often, those priorities are imposed by others. Even when the work plans are stable, clinical workers often contribute to several studies simultaneously and must balance their efforts appropriately between tasks. That requires not only a good understanding of the work assignments and flexibility, but also good judgment.

Strategic and Tactical Judgment

At the tactical level, each clinical worker contributes a particular expertise and performs a defined set of tasks in support of the clinical study or development program. In addition, there are important handoffs of work between coworkers: Some workers waiting for others to complete a task before they can start, and then handing their finished work to other workers who are waiting on them.

But clinical workers must also understand the strategic context of these tasks. By knowing the objectives of the study, the overall development program goal, the status of products being developed by competitor companies, and the business factors that are driving their efforts, they are in a much better position to plan their time wisely. They know how their work fits into the big picture and can make the right day-to-day decisions for the good of their team, the company, and the patients. Knowing the strategic goals also helps clinical workers to anticipate problems and mitigate risks.

Clinical workers are critical, analytical thinkers. They are accustomed to making sound, science-based decisions. However, they must also show good judgment when making operational decisions, following procedures, and interacting with other people. They must know the right thing to do, the right time to do it, and how to do it the right way, even when the work is difficult or unappealing. Sometimes, that means having the courage to "speak to power" and sometimes it is simply being diligent, complying with rules, and minding the details.

Rules and Details

Clinical studies of experimental drugs and medical devices are heavily regulated, because there are no shortcuts when it comes to human safety. Clinical workers have a keen sense of ethics and are committed to ensuring patient welfare. They know, accept, and comply with a multitude of regulations, processes, and procedures associated with patient care.

Regulatory agencies also set the data requirements for establishing the efficacy, safety, level of risk, and performance of a new medical product. Clinical workers realize that their work is pointless if their clinical study is flawed and the data they collect do not meet those regulatory requirements.

For both patient welfare and data integrity, the devil is in the details and there are lots of them. Precision, accuracy, completeness, and consistency are characteristic work habits of a good clinical worker.

Taking Initiative

Clinical studies are marathons not sprints. Good clinical workers pace their work, avoid distractions from their targeted goals, and persist despite unanticipated setbacks. They are highly motivated "self-starters" who can maintain their enthusiasm even when trying to solve difficult problems or accepting disappointing study results.

Developing a new drug or medical device means that the team is constantly exploring unknown scientific and medical frontiers. Clinical workers are continually grappling with medical conditions (and complications) that no one has seen before and for which there is no established course of action. Successful clinical teams work as a unit to adapt previous experiences to new situations, prepare for the unexpected, and creatively solve new problems. Good clinical team members therefore undergo a continual process of self-education. They have a natural curiosity and the ability to learn new concepts as they practice and push the boundaries of their craft.

2

How Are Drugs and Medical Devices Developed?

UNLIKE MANY CONSUMER PRODUCTS, new drugs and medical devices must be approved by government regulators before they can be sold. The manufacturer must demonstrate with convincing, hard data that the new medical product has been thoroughly tested and poses no unreasonable risks to the patients and healthcare providers who will be using it. Finding novel and innovative drugs and medical devices (i.e., research) is certainly a challenging and uncertain process for laboratory scientists and engineers. However, the longest, most expensive, and most demanding work is carried out by those who must collect the data on which market approval of the product is based (i.e., development). And of all the activities involved with product development, clinical studies receive the most attention and are arguably the most crucial. This book focuses on the employment opportunities and career potential of clinical study jobs.

OVERVIEW OF THE DEVELOPMENT PROCESS

Many diseases and medical conditions have unsatisfactory treatments and represent a wide range of opportunities for improving healthcare. The research and development (R&D) division, from which new medical products emerge, therefore performs an important function and represents a significant portion of a medical product company's workforce, facilities, and annual budget.

The R&D process begins with research scientists and engineers, who are charged with creating a new drug and medical device, respectively. They are given wide latitude in exploring scientific ideas, encouraged to exploit cutting-edge technologies, and allowed great flexibility in managing their time. Although they are expected to work toward the goal of identifying a useful new drug or medical device, they may make (and will be rewarded for) many contributions that are scientifically important but are only indirectly related to a marketable product. The probability of creating a new product that is commercially viable is relatively low. Some research scientists work in industry laboratories their entire career and publish many important scientific papers but retire without seeing a commercial product emerge from their efforts.

When the research scientists or engineers produce a product candidate that shows promise after laboratory testing, they hand off their work to specialists in preclinical development. The goal of the development scientists and engineers is to assess the product candidate's safety. Toxicologists conduct a standard battery of tests in animals and other biological systems to collect data on the toxic effects of drug candidates. Development engineers conduct well-controlled tests, sometimes using animals, to assess the risks of experimental medical devices. This preclinical development process usually takes 1–2 years.

If the results from the preclinical development tests show that the product is reasonably safe, the preclinical group hands the product to the clinical department for testing in patients. The major clinical steps involved in developing a new drug and medical device are illustrated in Figure 2-1. For drugs, clinical testing is broken into four phases. Medical device clinical studies are usually divided into two phases. Because the clinical data serve as the basis of the product's market approval, all clinical studies are designed and conducted according to the requirements imposed by regulatory agencies. (See the Drug Regulations and Medical Device Regulations sections of this chapter for details of these requirements.)

Drug Clinical Studies

The objective of Phase 1 clinical studies is assessment of the drug's safety in people. Before these studies, investigators only have results from the animal toxicology studies to estimate the drug's possible side effects. For this reason, the first Phase 1 studies are conducted very cautiously, with the drug used in small numbers of people, each of whom is watched very closely during drug administration. Typically, healthy adults who are willing to be confined and continuously observed are used in these studies, not patients. The data collected from these early studies determine the highest dose of the drug and the most frequent schedule of doses that can

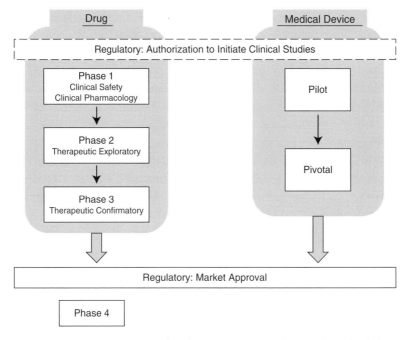

Figure 2-1. Major clinical steps for developing a new drug and medical device.

be safely administered. These early Phase 1 studies typically take 1–2 years to complete.

Another type of Phase 1 study is sometimes called a "clinical pharmacology" study. The objective of clinical pharmacology studies is to determine how human beings handle and dispose of the administered drug. To minimize complicating factors because of disease conditions, clinical pharmacology studies are also typically conducted in healthy adults. Various clinical pharmacology studies are scheduled to answer specific questions, such as how long it takes to excrete the drug; how the drug is metabolized; the effect of taking the drug under various dietary conditions (for example, before a meal or after a high-fat diet); and the comparability of different formulations of the same drug (for example, a tablet vs. a capsule). Clinical pharmacology studies are scheduled at different times throughout the clinical development program, and these data help the clinical teams decide how to design the studies that will be conducted with patients who suffer from the disease in question.

Phase 2 studies, which are sometimes called "therapeutic exploratory" studies, represent a shift from testing designed for safety assessment to testing that will provide the first evidence of the drug's therapeutic value. The first Phase 2 study marks the first time that the drug is used in patients who suffer from the disease that the drug is intended to prevent, diagnose, or treat. Although the Phase 1 studies will have provided some data outlining the drug's side effects, patients often respond differently than healthy people. For that reason, Phase 2 studies are well-controlled; patients are closely monitored for signs of adverse drug reactions, and drug treatment extends only for short periods of time. Phase 2 studies usually include only a few hundred patients per study. The number of Phase 2 studies and the length of exploratory drug testing vary depending on the type of drug and the nature of the disease being treated, but most of the data are collected in 2–3 years.

In exploring the therapeutic value of the drug, Phase 2 studies aim to determine the optimum dose level and dosing schedule, select the clinical end points that best reflect the drug's therapeutic effect (for example, a drop in blood pressure for an antihypertensive drug), rule in and rule out possible complicating factors (such as other prescription medications), and the appropriate patient population (for example, mild vs. severe disease). Because comparatively small numbers of patients are evaluated in Phase 2 studies, the data from these studies are usually insufficient to provide the definitive evidence of safety and efficacy required by regulatory agencies for market approval. However, these data provide valuable information for internal decision-making and planning discussions with regulatory authorities.

Almost all biopharmaceutical companies impose a formal Go–No Go decision point in their development program after the Phase 2 data have been collected. Their decision to proceed is based on the accumulated safety and efficacy data, the prevailing market and business conditions, and input from external reviewers including regulatory authorities. The Phase 3 clinical studies are by far the most expensive, and most important, portion of the development program. R&D senior managers want to make sure that the Phase 3 studies are performed right the first time. The data from the Phase 1 and Phase 2 studies are critically important, because they help the company's decision makers (1) in determining whether the drug has shown sufficient promise, in terms of safety and efficacy, to justify further investment, and, if so, (2) in guiding the design of the Phase 3 studies.

Biopharmaceutical companies may also hold discussions with external parties at the end of Phase 2, to assist them in making this important

Go-No Go decision. They may invite a panel of experts to provide an unbiased assessment of the data and solicit the experts' recommendations on the Phase 3 study designs. In addition, some regulatory agencies, such as the Food and Drug Administration (FDA) and the European Medicines Agency (EMEA), are willing to meet with biopharmaceutical companies, discuss their accumulated data, and comment on their plans for Phase 3 studies. This external input gives R&D managers greater confidence in making sound decisions regarding continuation, modification, or termination of their product development programs before starting a Phase 3 study.

The objective of Phase 3 studies, which are also called "therapeutic confirmatory" studies, is to confirm the preliminary evidence collected in the Phase 2 studies regarding the drug's safety and effectiveness in the intended disease and patient population. Although regulatory agencies will evaluate data from all of the studies when making a decision for market approval, data from the Phase 3 studies represent the most robust—and hopefully, the most compelling—evidence of the drug's true effects. Clinical workers therefore often call them pivotal studies.

In Phase 3 studies, the drug is administered to a significantly larger patient population (usually hundreds or thousands of patients) and tested under conditions closely resembling those under which it will be used after market approval. The larger patient group introduces greater diversity of population factors such as lifestyle, environmental conditions, and genetic traits, allowing investigators to assess the drug's therapeutic value and side effects more realistically. However, although this patient group is meant to be representative of the general patient population, the investigators restrict the variability in the patient's disease characteristics. For example, the study may restrict enrollment to only lung cancer patients (and not any other type of cancer), but permit lung cancer patients of any age, occupation, or nationality. Because patients may be treated and evaluated for a long period and the patient population is large, Phase 3 studies usually take 3–5 years to complete.

Whereas research scientists explore many novel ideas and test thousands of drugs to find one suitable for development, clinical workers test and reject about 20 drugs for each one that receives market approval. Clinical development plans can be derailed by poor test results, technical difficulties, changes in the marketplace, or new regulations. Some of these difficulties can be overcome by additional testing or redirection of the development strategy. In other cases, the only alternative is to stop development of the drug in question and turn to a more promising drug candidate.

For drugs that are successful, assessment of a new drug often extends beyond market approval. Clinical studies conducted after market approval are called Phase 4 studies and are used to collect additional data on the drug. For example, if the drug company wants to sell a liquid formulation of a drug that was already approved as a tablet, it must collect data in a Phase 4 study using the liquid form. If the drug was approved to treat arthritis, the company would need to conduct Phase 4 studies to show that the drug is also effective in treating psoriasis before it can be marketed for that disease. Phase 4 studies are often larger and may be less rigorously controlled than the pivotal Phase 3 studies. However, they must meet the same regulatory standards for patient welfare and data integrity as all other clinical studies.

Medical Device Clinical Studies

Clinical studies on medical devices are typically divided into two phases: pilot and pivotal studies. The objective of pilot studies is to gain initial clinical experience with the device under conditions that minimize the risks to patients. The size and length of pilot studies vary, depending on potential risk to the patient; higher-risk devices are initially tested in a relatively small number of patients. Pilot-study patients are monitored closely for signs of side effects, and the length of follow up is proportional to the potential risks. Investigators gather data on the short-term safety of the device, evaluate treatment technique(s), and explore the optimal patient population. Manufacturers also conduct pilot studies to explore potential new uses for a medical device beyond its originally intended use.

Often, research engineers use the results of pilot studies to refine the design, materials, and performance characteristics of their medical device. Although these engineering changes usually give the device advantages over the original product, the new medical device may have changed enough that it may also need to be assessed in pilot studies. If the changes have been significant, it is considered a new product, and the clinical data obtained from the earlier model will not support its market approval. A series of such engineering refinements and pilot studies may be required to optimize the medical device's performance.

When engineering issues are no longer a concern, investigators conduct one or more pilot studies to assess the feasibility of conducting a pivotal study. The results from these pilot studies will help refine the design of the pivotal study, identify the most suitable study end points, estimate the magnitude of the response to treatment, and assess the variability in patient responses. In an appropriately designed pilot study, the number of enrolled patients and the duration of patient follow up are sufficient to provide data that will justify initiating the pivotal study.

The pivotal study provides clinical data that forms the basis of marketing approval of the medical device. Therefore, the study must collect enough data under the appropriate test conditions to establish the device's safety and effectiveness. Drawing from their experiences with the small pilot studies, investigators design the pivotal study to assess the device's performance under conditions that are appropriately controlled and minimize bias in interpreting the results. To ensure that the study patients are representative of the general patient population, pivotal studies usually involve a wide variety of patients at multiple clinical sites. The duration of pivotal studies varies, depending on the type of medical device. An implanted device such as an artificial hip requires a longer follow-up period than a single-use device such as a nicotine patch.

For medical devices that reach the clinical stage of testing, the chances for success are relatively high, because serious safety issues are usually addressed through extensive animal and laboratory testing. Even in situations in which engineering changes necessitate a number of pilot studies, the clinical program for a medical device is usually much shorter and requires a smaller total clinical dataset than that required for a new drug. However, if the product combines features of a drug and medical device, such as an asthma inhaler or an insulin pump, the clinical testing program is almost as extensive as for a conventional drug.

REGULATORY CONSIDERATIONS

Although various initiatives have succeeded in harmonizing the regulatory requirements of various countries for market approval of drugs and medical devices, many differences and variations remain between countries. This creates a challenge for both the clinical teams, which often conduct their clinical studies in multiple countries, and for the company's executives, whose goal is to obtain multinational market approval of their new product as quickly as possible.

To gain efficiencies throughout the long process of collecting and compiling clinical data, therefore, many companies opt to follow the regulatory requirements of the most stringent agency. Their rationale is that by meeting the tough regulatory requirements of this agency, the company will likely be able to satisfy the regulatory requirements of other countries as well. The United States is considered a large and lucrative market for medical products, and most companies include the United States in their marketing objectives. Therefore, the FDA quite often represents the regulatory standard that dictates the company's decisions for designing and conducting clinical studies.

Drug Regulations

The regulatory agencies of many countries in North America, Europe, and the Asia-Pacific rim have adopted regulations specified by the International Conference on Harmonization (ICH) for development and market approval of new drugs. Although the agencies in individual countries may impose additional or modified regulations, the ICH guidelines have permitted drug companies to standardize their work in a way that will likely be acceptable around the world. The ICH regulations associated with drug development fall into three categories: product manufacturing, animal studies, and human studies.

The chemists, molecular biologists, and other scientists who make the drug provide a vitally important R&D service. Whether they produce small drug batches in the laboratory or kiloton quantities of commercial product, no experimental studies can be conducted and no patients can be treated unless they do their job—and do it well. Once the research scientists have produced a drug candidate worthy of development, it falls to their colleagues in the manufacturing division of R&D to make sufficient quantities of the drug, of appropriate quality and at the right time, for testing in animals and people.

Drugs must meet high standards for purity, chemical composition, and quality, and those who produce the drug must perform their work under carefully controlled conditions and maintain meticulous records. The ICH standards for drug quality are called Good Manufacturing Practices (GMPs) and go into considerable detail regarding requirements for the production, equipment cleanliness, packaging, storage, and quality testing of each drug batch.

Research scientists who conduct only exploratory studies in animals and other biological systems have considerable freedom in how they perform and document their studies. Virtually their only restrictions are the general animal welfare provisions imposed by animal protection laws and the record keeping needed to support patent applications.

However, when toxicologists begin conducting animal studies and collecting data that will be used to characterize an experimental drug's safety, they must comply with ICH standards and Good Laboratory Practices (GLPs). ICH specifies detailed requirements for the design of toxicology studies: choice of species, dose levels, number of doses, length of drug treatment, data to be collected, analysis methods, and content of the final report. Studies in the battery of ICH-required toxicology assessments are scheduled to coincide with the conduct of the clinical studies that they support.

GLP regulations set standards for the organization, personnel qualifications, physical structure, maintenance, and operating procedures of the toxicology laboratory's testing facilities. By complying with these GLP regulations, development scientists can assure the regulatory authorities that their toxicology studies met the regulatory standards for quality and data integrity.

All of the data confirming the quality and purity of the drug batches and the data showing the drug's effects in animals are collected for one purpose: to establish that the drug will not expose humans to unreasonable risks when used in clinical studies. This evidence is formally submitted to regulatory authorities who determine whether the data are sufficient to justify exposure of the drug to people.

In the United States, companies submit an Investigational New Drug (IND) application, which contains the accumulated data from their animal studies, information describing the drug's chemical characteristics and production, and the protocol for the first proposed clinical study. FDA reviewers have 30 days to assess the information in the IND and can prevent the clinical study from starting if they have concerns; if not, the company may proceed with its study. The FDA also assigns a tracking number to the IND application. As the company submits new data and conducts new clinical studies with the drug, it references this tracking number.

Member countries in the European Union (EU) use a consolidated procedure for regulating clinical studies. When a company wishes to conduct a clinical study in one or more member countries, it submits a Clinical Trial Authorization (CTA) application to the appropriate regulatory authority. EU member countries have established differing procedures for reviewing and authorizing CTA approvals, but the requirements for data and other associated documentation are similar to IND requirements. The company is also responsible for obtaining a EudraCT number, which is maintained in a European database and used to track the clinical study, whether the study is conducted in one or several member countries. Unlike the IND tracking number and filing system, which applies to all clinical and nonclinical studies associated with one experimental drug, a unique EudraCT number is assigned for each clinical study.

To initiate clinical studies with a new drug in any country, the regulatory authorities only accept GLP-compliant laboratory data as valid evidence of the drug's preclinical safety and will permit only GMP-compliant drug supplies to be administered to people. These data and quality considerations are also important to another group that oversees patient welfare. In some countries including the United States, the group is called the institutional review board (IRB); in other countries it is an independent ethics committee (IEC).

IRBs/IECs typically operate under the auspices of an individual research institution; however, some operate independently within a geographic region. These groups consist of at least five members who have the qualifications and experience to evaluate the science, medical aspects, and ethics of proposed clinical studies. One member must be a nonscientist and typically represents patient advocacy groups or is a member of the clergy. If the IRB/IEC is associated with a research institution, one member must be independent of that institution.

The IRB/IEC reviews the clinical protocol, investigator credentials, and other relevant documents to determine whether the rights, safety, and well-being of the patients will be protected. The study may proceed at the institution(s) under its jurisdiction only with its approval. The IRB/IEC may also terminate or suspend an ongoing study to protect the rights, safety, or well-being of the study patients.

Good Clinical Practices (GCPs) issued by the ICH is the main regulatory standard for clinical study conduct. It specifies the responsibilities and tasks of the company, clinical investigators, and IRB/IEC for protecting patients, handling data, record keeping, reporting adverse events, handling experimental drugs, and study monitoring. GCPs also specify the information that must be contained in the study protocol, investigator's brochure, and informed consent form.

For clinical studies that involve treating patients who are gravely ill or experimental drugs that are potentially hazardous, sponsors may establish an independent data monitoring committee (DMC). The DMC consists of clinicians with expertise in the clinical area under study and at least one biostatistician. Unlike the IRBs/IECs, the DMC periodically assesses the progress of the study, the accumulated safety data, and sometimes selected data regarding the drug's therapeutic benefit. Although the drug company pays DMC members for their efforts, the DMC operates independently from the company and makes recommendations whether to continue, modify, or stop the study. The DMC recommendations are based not only on concerns for patient welfare but also on whether the data will yield conclusive results. The DMC may feel that it is futile to continue a study that is accumulating highly variable and inconsistent data.

When sufficient clinical data have been collected from the Phase 1, 2, and 3 clinical studies, the company applies for market approval. ICH defines a standard format for organizing the manufacturing, animal, and clinical data called the common technical document (CTD). Although a separate marketing authorization application (MAA) must be submitted to each country in which the company seeks to sell the drug, the CTD format allows the company to compile this massive documentation efficiently and submit it to multiple countries with only minor adjustments. In the United States, for example, the MAA is called a New Drug Application (NDA), but the requirements for content are harmonized with the CTD format. In Europe, the EMEA allows companies to submit an MAA via a centralized procedure, which grants market authorization in all EU countries.

Medical Device Regulations

International standards for managing the quality and risk associated with medical devices have been established by the International Organization of Standardization (ISO). In addition, individual countries have established requirements to regulate medical devices. In the United States, the FDA oversees the regulation of both drugs and medical devices. In Europe, the European Commission sets common medical device regulatory standards that apply to all member countries in the EU, but each country has procedures for product approval. Because of the diversity and innovativeness of medical device types, those regulations cover a wide range and depend on the device's intended use and relative risk.

Class I and most Class II medical devices pose minimal risk to patients, and manufacturers can provide sufficient evidence of safety and performance without conducting clinical studies. Regulatory agencies typically grant market approval of these devices on the basis of documentation describing the product manufacturing specifications, design considerations, materials used in production, and the intended use of the device.

For Class III medical devices, such manufacturing evidence is insufficient to demonstrate the safety and effectiveness of the product. The manufacturer must also provide data from clinical studies. The objective of the clinical studies is to verify that under normal conditions of use the medical device performs as claimed by the manufacturer, identify undesirable side effects, and assess whether the risks are reasonable when compared with the device's intended use. ISO, the European Commission, and the FDA have all established regulations governing the clinical evaluation of medical devices.

As with clinical investigations of drugs, clinical evaluations of medical devices are overseen by IRBs/IECs and must comply with GCPs. In the United States, companies must obtain an investigational device exemption (IDE) from the FDA before initiating clinical studies with Class III devices.

For market approval of a Class III medical device in the United States, the company must submit a Premarket Approval (PMA) application to the FDA. Companies wishing to obtain market approval in Europe must meet the requirements of the appropriate Council Directive (issued by the European Commission) governing the respective type of medical device. The regulatory authorities base their market approval on a determination that the application contains valid scientific evidence that the device is safe and effective for its intended use. One pivotal clinical study, if it has been appropriately designed, usually provides sufficient evidence of safety and effectiveness. The manufacturer must also submit extensive documentation regarding the device's design, engineering features, and commercial manufacturing process. Unlike drug applications, the portions of a medical device marketing application dealing with manufacturing issues often are larger than the sections describing clinical studies.

INDEPENDENT CLINICAL STUDIES

Biopharmaceutical and medical device companies are not the only ones who conduct clinical studies. Individuals with the appropriate medical qualifications often conduct clinical studies as part of their basic research activities, such as clinicians at academic medical centers, private research institutes, and government hospitals such as the National Institutes of Health and the Veterans Administration. Typically, these "independent" clinical studies are not conducted to evaluate an experimental drug or device, because independent investigators rarely have access to products that meet the GMP and ISO requirements for purity and risk mitigation, respectively. Instead, independent clinical studies address other medical questions, such as testing a new surgical procedure, assaying the genetic mutations in tumors, comparing brain scans of healthy people versus mentally impaired patients, or testing the effect of a marketed drug in a new patient population.

Independent clinical studies must comply with the same rules regarding patient welfare as commercial R&D organizations. The clinical protocol must have a defined, scientifically valid objective and be reviewed and approved by an appropriate IRB/IEC. The investigator must also comply with GCPs, including voluntary, informed consent of the people who participate in the study. If the clinical study includes the use of a commercially available drug or medical device, the investigator must also comply with all applicable regulatory requirements, such as submitting an IND, IDE, or CTA application and reporting adverse reactions during treatment.

THE STAGES OF A CLINICAL STUDY

The fascination and excitement of clinical studies comes from testing a new medical product in patients and assessing how well it works. Each experimental product and each disease presents unique challenges and opportunities. Understanding the science behind the disease and experimental product makes each new study fascinating. Improving the condition of patients with serious health problems, despite technical and logistical challenges, is always gratifying. All of this variety in diseases, product characteristics, and patients guarantees that no two clinical studies are ever exactly alike.

However, the standards for designing and conducting a clinical study and safeguarding the patients' welfare are specified by the GCPs and are the same, regardless of the medical product, the disease, or the goal of the study. Figure 2-2 shows the GCP-mandated stages of a typical clinical study.

A team of clinical specialists is responsible for planning, conducting, and completing each clinical study. In the rest of this book, you will learn about these clinical specialties through the work of a typical clinical study team at a typical medical products company. The team members and their company, CanDo Pharma, are fictional, but their work is characteristic of that found at all biopharmaceutical, medical device, and contract research organization (CRO) companies. To simplify the examples presented in the following chapters, the CanDo Pharma team will be testing a drug in its clinical study; differences in procedures, requirements, and practices for medical device studies will be noted in the chapters, when applicable.

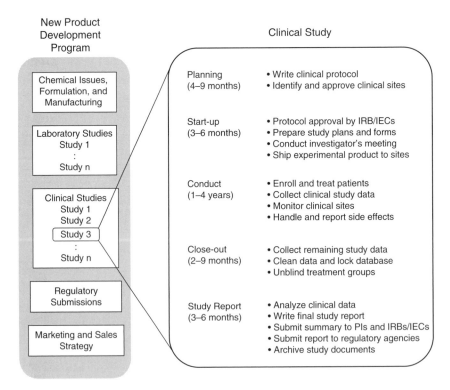

Figure 2-2. GCP-mandated stages of a typical clinical study.

In general, learning how to conduct clinical studies of drugs, whether at a biopharmaceutical company or a CRO, is a good way to start your clinical career. The regulatory requirements governing drug studies are more rigorous and more extensive than those governing medical device studies. In addition, the number and variety of jobs needed to support drug studies are greater than the openings for medical device company positions. Those with drug study experience can easily transition to medical device jobs, but those with only medical device experience often find it difficult to meet the requirements for clinical study positions at drug companies.

A clinical study is one component of the overall development program for a new medical product. In a typical program for a drug, a development strategy team oversees dozens of clinical studies; development programs for Class III medical devices usually include a handful of pilot studies and one pivotal study. In either case, a medical director usually leads the development strategy team and is assisted by a project manager. In our example, Dr. Abernathy is the medical director. He is the strategy team leader and the physician responsible for medical oversight of the CanDo Pharma drug studies. Bill, a project manager, works closely with Dr. Abernathy to manage the development program's timelines, milestones, and resources. (Chapter 11 describes more fully the role of the project manager.)

In conjunction with other senior members of CanDo Pharma's strategy team, Dr. Abernathy defines the list of clinical studies and establishes each study's goals. To implement each clinical study, Dr. Abernathy's strategy team establishes a clinical study team, which prepares the protocol and is responsible for overseeing the study. The study protocol is a document that transforms the strategy team's goals into a detailed plan stating the study's design, criteria for selecting patients, treatment procedures, clinical assessments, and data analysis methods. (In medical device studies, the protocol is called a "clinical plan." The main objective of the clinical plan is to outline how the study will document and evaluate the risks associated with normal use of the medical device.)

While writing the protocol, the study team identifies one or more clinical sites and makes arrangements for conducting the study. Biopharmaceutical companies such as CanDo Pharma and medical device companies do not conduct clinical studies themselves, but rather engage independent physicians to do the studies on their behalf. The CanDo Pharma team chooses physicians who have the appropriate medical expertise and selects locations that are likely to have large patient populations with the target disease or medical condition.

CanDo Pharma pays these independent physicians for their work under a contract. In return, the physicians promise to follow the protocol and comply with regulations for protecting the patients and the integrity of the clinical data. CanDo Pharma is called the study sponsor, because the company pays for the study and supplies the experimental drug. The independent physicians who agree to conduct the study are called principal investigators (PIs), and the locations where the study is performed are called investigator, or clinical sites. (The activities and responsibilities of a clinical site are detailed in Chapter 3.)

After the study team members finish the protocol and finalize the clinical site contracts, they complete a number of start-up tasks. For example, they submit the protocol to government regulatory agencies. (In the United States, the FDA must receive a copy, referencing the assigned IND number. In Europe, the protocol is assigned a EudraCT before submission to the appropriate regulatory bodies.) In addition, the PIs at each clinical site submit the protocol to their local IRB (or IEC). The regulatory authorities and IRBs review the protocol to ensure that the study is scientifically sound and that the patients will be treated ethically.

At the same time, the team completes several other important activities, such as writing detailed plans and designing customized forms. These documents are needed to collect and process the data from the study as specified in the protocol. (The following chapters explain these activities and documents in greater detail.)

When the team has successfully completed all preparations for the study, Dr. Abernathy schedules an investigator meeting for the participating PIs and their key assistants. The purpose of the investigator meeting is to make certain everyone thoroughly understands the protocol, to train the site staff on unfamiliar study procedures, and to review how to collect and report the study data. Following the investigator meeting, the CanDo Pharma team ships the experimental drug to each clinical site and authorizes the PIs to start the study.

Regulatory agencies require CanDo Pharma to post certain details of the protocol and investigator sites on public-access websites. This allows healthcare professionals, interested patients, and the general public to learn about the study.

During study execution (also called study conduct), the PIs at each site begin enrolling and treating patients. Each patient who is interested in participating must meet the medical requirements specified in the protocol. The patients who qualify and agree to participate receive the experimental drug in doses according to the protocol's schedule.

The study team conceals the identity of the CanDo Pharma drug using a coded packaging system. Neither the patients, the PIs, nor the study team at CanDo Pharma know which patients receive the experimental drug and which receive a placebo. (Placebos are blank treatments that look identical to but contain no test drug.) The process of concealing the identity of the experimental drug and placebo is called "blinding." By blinding their observations, investigators collect and assess the data fairly and without bias.

The PIs conduct and record the results of laboratory tests and other clinical procedures on the patients before, during, and after completion of the drug treatment schedule. Some studies are short; others last for years. Especially for long studies, the CanDo Pharma team periodically collects and reviews the accumulated, but blinded, data. If any of the required data are missing or appear to be illogical (for example, a man ten feet tall), the CanDo Pharma team will ask the PI to explain or clarify the apparent error.

During the course of the study, the study team at CanDo Pharma and the clinical sites maintain a close working relationship. The CanDo Pharma team regularly contacts the PIs to check on their progress and periodically visits each clinical site to oversee the study first-hand. These discussions and visits ensure that the PIs are administering all protocol-specified treatments and procedures properly, recording the data accurately and completely, and handling unanticipated problems regarding patient safety and data integrity.

Dr. Abernathy carefully reviews complaints and discomfort reported by the patients. Side effects experienced by a patient may justify breaking the code to determine which treatment the patient received. Side effects that are unexpected or unusual may require both medical and regulatory follow-up for the patient's safety.

After the PIs collect the last data points on the last patient treated according to the protocol's schedule, the study team formally closes the study. The team reviews the study database one last time for accuracy and resolves any missing data. This process may include collecting additional information from the clinical sites to answer lingering questions about their data entries. The study team then closes the database. The blinding code is broken to reveal which patients received drug versus placebo, and the team analyzes the data according to a predefined statistical analysis plan.

The decoded and analyzed results allow Dr. Abernathy and the other team members to draw conclusions about how well the experimental drug worked in the enrolled patients. They document these results and conclusions in a formal clinical study report. Once Dr. Abernathy and other key team members sign the clinical study report, CanDo Pharma formally submits it to the appropriate regulatory agencies (for example, the FDA in the United States). CanDo Pharma also complies with regulatory requirements for archiving the study documents and posting the key study results on public-access websites.

THE STUDY TEAM MEMBERS AND THEIR RESPONSIBILITIES

A number of people work under Dr. Abernathy to conduct CanDo Pharma's clinical study. The study team members introduced and profiled throughout this book are fictional, but they represent composites of the backgrounds, training, skills, work habits, and responsibilities that you would expect to find in actual team members who perform these jobs at a real company. Whether the person works at a sponsor company (biopharmaceutical or medical device) or at a CRO, the roles and responsibilities are generally the same. If specific job responsibilities differ between these industry segments, the chapters include explanations of the differences.

Figure 2-3 shows the various clinical specialties that support the clinical development program and that are represented on the clinical study team. Next to Dr. Abernathy, the clinical study manager carries the most responsibility for the clinical study's success. Nancy is the study manager on the CanDo Pharma clinical study team and has overall responsibility for planning, overseeing, and completing the study correctly and on time. All the other team members, whether they report to her or not, are accountable to Nancy for their work on the study. Chapter 11 describes the clinical study manager's role in detail.

One or more clinical research associates (CRAs) assist the clinical study manager with planning, starting up, and monitoring the study. David is a CRA on Nancy's clinical study. From his desk at CanDo Pharma, David works under Nancy's supervision and handles the day-to-day study tasks. Much of the study's success depends on his ability to

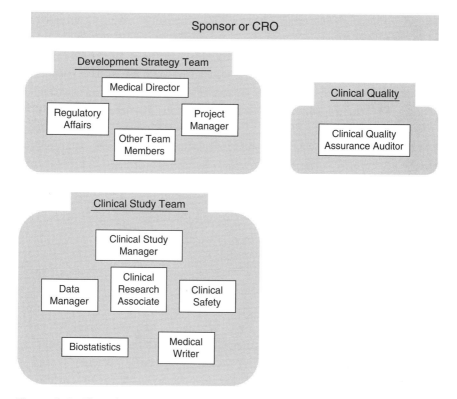

Figure 2-3. Clinical specialties that support the clinical development program and that are represented on the clinical study team.

maintain productive communications between the investigator sites and the study team, and he makes regular visits to the sites to monitor their activities. Chapter 4 fully describes CRA responsibilities.

A data manager handles the clinical data collected during the study. Maria is the data manager on Nancy's study team, and her main responsibilities are preparing the case report forms and managing the study's clinical database. The case report forms are equivalent to a laboratory notebook and are used by the clinical sites to record all of the data required by the protocol. A clinical database is programmed computer software, which organizes and stores all of the study's data for later analysis. Chapter 5 gives a complete description of data manager responsibilities.

The clinical safety specialist assigned to Nancy's study team is Jasmine. Her primary responsibility is to monitor the study results for signs of potential drug safety issues. If the side effects experienced by study patients meet certain criteria, Jasmine must compile an adverse event report and forward it to the regulatory authorities. Chapter 9 elaborates on clinical safety positions.

A biostatistician handles all statistical aspects of the study. Tom is the biostatistician on Nancy's study team. Before the protocol is written, he advises the team on the statistical considerations for designing the study. Tom contributes to the study protocol by describing the statistical procedures he will use during data collection and analysis. He also analyzes the final study results. Chapter 6 provides details of biostatistician positions.

Mike is the medical writer on Nancy's study team. With input from the study team, Mike writes the study protocol and other official documents required by regulatory agencies throughout the course of the study. At the end of the clinical study, he writes the study report in a format that follows CanDo Pharma's standards and is easy for the regulatory authorities to read. (See Chapter 10 for further information about medical writer positions.)

Throughout the study, Nancy's study team must carefully comply with the requirements imposed by government regulatory agencies. Amy, a regulatory affairs specialist, manages all regulatory issues associated with the CanDo Pharma drug. Because the study will be conducted at sites in the United States, Australia, and Germany, Amy advises the study team regarding the regulatory requirements of all three countries. The team channels all communications with the regulatory authorities through her. Chapter 8 provides a detailed description of jobs in regulatory affairs.

Clinical quality assurance (CQA) specialists monitor all product development activities to ensure that the clinical teams follow regulatory requirements and CanDo Pharma's own standard operating procedures (SOPs). CanDo Pharma established its own CQA department, but some companies use the CQA services of a CRO. Because the CQA department is organizationally separate from the clinical study team and from Dr. Abernathy's strategy team, it provides independent quality oversight. CQA auditors randomly select and audit a portion of CanDo Pharma's clinical studies. If Nancy's study is chosen, a CQA auditor, such as Carlo, conducts an internal audit. Carlo reviews the team's activities and documents during his audit and records his findings in an audit report. (For additional information about CQA positions, see Chapter 7.)

In the following chapters, you will learn more about the members of Nancy's clinical team, how they got their jobs, the details of their daily work, and the kinds of career advancement they can expect after mastering their current responsibilities. In addition, Chapter 3 describes the work at a typical clinical site and, importantly, the role of the study coordinator. Sally is one of the study coordinators participating in the CanDo Pharma study. Through Sally, you will learn how study coordinators get their jobs and how the clinical experience they gain helps them qualify for entry-level clinical positions in industry.

CLINICAL DEVELOPMENT RESOURCES

Agency Standards

International Conference on Harmonization (www.ich.org) sets international standards for clinical studies of investigational drugs.

International Organization of Standardization (www.iso.org) sets international quality standards recognized by governments for commercial products.

    ISO 14155: General requirements for medical devices in clinical studies

    ISO 14971: Risk management of medical devices

    ISO 13485: Quality management of medical devices for regulatory purposes

Food and Drug Administration (www.fda.gov) sets safety regulations for foods, drugs, and other medical products in the United States.

European Medicines Agency (www.ema.europa.eu) evaluates and supervises safety regulations of drugs in the European Union.

European Commission/medical devices (http://ec.europa.eu/enterprise/sectors/medical-devices/index_en.htm) sets regulatory standards for medical devices in the European Union.

Good Clinical Practices Training and Certificate Programs

ClinfoSource (www.clinfosource.com) offers online GCP and product development courses and a certificate program in clinical site activities for CME and CNE credit at reasonable cost.

Kriger Research Center International (www.krigerinternational.com) offers online GCP courses and certification in multiple languages on the country-specific web pages under the "Courses" tab.

Medical Research Management (www.cra-training.com) offers online and classroom GCP training for ACPE credit.

Association of Clinical Research Professionals (www.acrpnet.org) offers online GCP courses and certification for CME, CNE, and ACRP credit at reasonable cost.

GCP Training Online (www.gcptraining.org.uk) offers online GCP training and certification, emphasizing the European Clinical Trial Directive, at reasonable cost.

Institute of Clinical Research (www.instituteofclinicalresearch.com) offers online GCP and European Clinical Trial Directive courses and certification at reasonable cost.

3

What Happens at the Clinical Site?

A S MENTIONED IN CHAPTER 2, sponsors and contract research organizations (CROs) typically do not conduct clinical studies within their companies. They engage physicians who are not employees of the sponsoring company and who are located at facilities that are not owned or operated by the sponsor. Why do sponsors willingly rely on "outsiders" to recruit, treat, and evaluate the patients in their study?

First, regulatory agencies require that clinical studies of experimental products must be supervised by qualified physicians. No drug or medical device company can afford to hire and maintain medical experts in all the therapeutic areas covered by their clinical studies. Instead, when they need specific expertise, they engage independent medical specialists who have active medical practices. Frequently, sponsors select top-notch physicians who are not only specialists but also recognized as influential "opinion leaders" in their field. Although these independent physicians are paid for their time and reimbursed for their expenses, sponsors take precautions to ensure that the principal investigators (PIs) remain objective. The precautions include enforcing patient enrollment rules, masking the study treatments, using third party experts to interpret qualitative data such as X-rays, and requiring disclosure of any financial interest the PI may have in the sponsor's business.

Second, sponsors and CROs choose physicians and locations that can supply a sufficient number of qualified patients. The clinical sites may be university medical centers, hospitals, or the private practices of physicians. In addition to having the proper medical qualifications, physicians at these locations have access to patients through their own clinic or a nearby medical center. Some clinics may be devoted entirely to treatment of one disease (e.g., cancer centers); physicians at specialized centers are

43

usually very interested in conducting clinical studies, which offer new and otherwise unavailable treatment options for their patients.

Third, sponsors and CROs seek clinical sites that have appropriately equipped facilities. Clinical studies often require specialized medical procedures or equipment that may not be available everywhere. Because the requirements vary greatly from one study to the next, sponsors cannot build and efficiently maintain all of these capabilities within their companies.

Finally, regulatory authorities stipulate that sponsors must not have direct contact with the patients who participate in their clinical studies. Sponsors must design studies so that they cannot trace the study results to individual patients. Working with independent physicians makes it easier for sponsors to safeguard patient confidentiality. Because PIs maintain detailed medical records and are responsible for treating and caring for the enrolled patients, they clearly know the identity of their patients. However, all information that the PIs forward to the sponsor must be coded to conceal the patient's identity.

Phase 1 studies are the only exception to this separation between sponsor and study site. Because the purpose of Phase 1 studies is safety and healthy volunteers are commonly used instead of patients, the issues regarding confidentiality are somewhat different. Some big pharmaceutical companies and a number of large CROs maintain their own Phase 1 units, which have special facilities and a trained medical staff for conducting Phase 1 studies. (See Chapter 2 for more information about Phase 1 studies.)

ACTIVITIES AT THE CLINICAL SITE

Figure 3-1 illustrates the roles and relationships of those who work at a clinical site. Among the PIs that the CanDo Pharma team selects for its clinical study is Dr. Chase, a member of a group medical practice in a large city in the United States. As a PI, Dr. Chase takes overall responsibility for conducting the CanDo Pharma study at her location, including the care of all patients who participate at her site. She has conducted previous clinical studies and has all the facilities required by the CanDo Pharma protocol. Dr. Chase is also confident that a number of her patients will be interested in participating in the study.

Because of Dr. Chase's experience in conducting clinical studies, CanDo Pharma invited her to comment on the draft protocol. PIs often

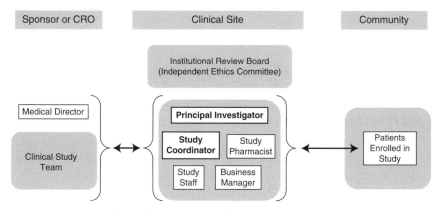

Figure 3-1. The roles and relationships of personnel who work at a clinical site.

spot difficulties with procedures that are inconvenient or impractical for the patient or site staff. For example, Dr. Chase suggested increasing the time interval between two of the study procedures, because they must be performed in opposite wings of the hospital. She also suggested changing the treatment schedule, which required patients to take one of their drug doses at 2 AM each morning.

Company sponsors pay the PIs for their time, their research staff's salaries, and the cost of all special procedures that are required by the protocol. CanDo Pharma's contract defines Dr. Chase's obligations as a PI and the terms for paying her for her work. Dr. Chase's medical group has an office manager, who manages the study contract and keeps track of the payments they receive from CanDo Pharma. At larger institutions such as a university medical center, a separate business office typically handles the budget and contract work on behalf of the institution's PIs.

Dr. Chase and her staff conduct all clinical study activities that involve patients. These include:

- Recruiting: identifying or soliciting patients

- Screening: assessing interested patients to determine whether they qualify

- Consenting: obtaining written agreement from the patient to participate

- Enrolling: adding qualified patients to the study

- Randomizing: randomly assigning patients to one of the study's treatment groups

- Treating: administering the experimental treatment (drug or placebo)
- Evaluating: conducting the study procedures and collecting the data
- Recording: entering the study data on case report forms
- Archiving: storing all patient records after closing the study

The relationship between the sponsor and the clinical sites is complementary and interactive. Dr. Abernathy knows all about CanDo Pharma's drug but must rely on Dr. Chase (and the other PIs) to treat the patients and collect the study data. Dr. Chase knows all about her patients' medical conditions but must rely on Dr. Abernathy to supply CanDo Pharma's drug and instructions for its use.

The CanDo Pharma team prepares an investigator's brochure (IB), which provides Dr. Chase with complete information about CanDo Pharma's drug. The IB describes the physical and chemical properties of the drug; summarizes the effects of the drug in animal studies; summarizes the accumulated data from previous studies in humans, both healthy volunteers and patients; provides guidance on how to administer the drug; and describes precautions and recommended treatments for anticipated side effects. In addition to the IB, though, ongoing communication between the sponsor and the clinical sites is essential to share information and make informed decisions, especially when problems arise. The sponsor and PIs work jointly, within the boundaries imposed by the regulatory authorities, to safeguard the patients' welfare and conduct the study properly.

THE ROLE OF THE STUDY COORDINATOR

The study coordinator is the person responsible for overseeing all the day-to-day tasks required by the clinical study protocol at an investigator site. Sally is Dr. Chase's study coordinator. Under Dr. Chase's direction, Sally communicates with the local institutional review board (IRB), recruits and screens patients, schedules patient visits during the study, assists with study procedures, records the study data on the official case report forms (CRFs), and communicates with CanDo Pharma throughout the study. Sally is also an important resource for the office manager regarding the study's budget and contract.

BOX 3-1. *Study Coordinator Responsibilities*

- Follow study protocols and GCP requirements
- Set up study files
- Submit IRB/IEC documents
- Assist with training site staff
- Screen and enroll patients
- Schedule study procedures
- Oversee inventory and storage of experimental drug or medical device
- Maintain patient charts and source documents
- Complete case report forms
- Respond to CRF queries
- Report adverse events
- Complete sponsor surveys
- Prepare study budgets
- Store study records after study completion

Landing the Study

For clinical sites, the bidding for studies can be very competitive. Often, sponsors approach selected PIs, because they know and respect his or her reputation or the clinical site has performed well on the company's previous clinical studies; other PIs who want to participate may be excluded. The PI may be aware via published literature or medical conferences of a product under development, has patients who may benefit from such treatment, and asks to be included as a study site.

Sally was instrumental in helping Dr. Chase to clinch the study sponsored by CanDo Pharma. Sally completed a CanDo Pharma survey, which asked her to describe the site's facilities, staffing, and performance on previous clinical studies. Sponsors use surveys as a first step in assessing the suitability of a site. Although the PI's site may have successfully participated on a previous study, it may not be acceptable on a new study with different objectives and requirements. All investigator sites aim to meet the regulatory requirements regarding the staff's credentials, security for storing experimental products, and IRB oversight. The more variable, but certainly most important, factor in site selection is the availability of patients who meet the protocol's criteria for enrolling in the study.

Sally plowed through the office's medical records and checked with colleagues in Dr. Chase's group practice to estimate how many of their patients might qualify for the study. She also filled out a table in the CanDo Pharma survey, giving detailed information on Dr. Chase's success in recruiting qualified patients on previous studies. Although the CanDo Pharma study is different than those studies, previous success in recruiting and retaining patients is often an indicator that the PI will be successful on a new study.

Sally worked with her institution's business manager to draft a study budget, based on the work specified in the protocol. Each investigator site sets its own fees for the cost of treatment, clinical procedures, and staff services required by the protocol, as well as the overhead rates to cover administrative costs. Although the study budget is negotiable, CanDo Pharma, like most sponsors, usually accepts the budgets that are proposed by high-performing and desirable clinical sites. Sites that are forced to cut costs might compromise the integrity of the clinical data or patient welfare, neither of which is acceptable to the sponsor or regulatory agencies.

Sometimes, despite being fully qualified and highly desired by the sponsor, the site may not wish to participate in the study. Knowing this lack of enthusiasm early, via the site's responses to questions on the survey, allows the sponsor to concentrate its site selection efforts elsewhere. Because Dr. Chase was keen to participate in the CanDo Pharma study, Sally's answers on the survey conveyed their high level of interest.

For sites that submit a favorable survey, the next step in the selection process is an on-site inspection by the sponsor or a designated CRO. Sally cooperates with David, CanDo Pharma's clinical research associate (CRA), to make the site evaluation visit productive. Together, they walk through the areas where the patients would be treated, the experimental drug would be stored, and the blood and other laboratory samples would be processed. Sally also provides copies of documents required by the regulatory authorities, such as the list of IRB members and résumés for all of Dr. Chase's clinical study staff. (See Chapter 4 for details of site evaluation activities.)

Study Preparation

After CanDo Pharma selects Dr. Chase's site and negotiates the study budget and contract, Sally assembles the documents for IRB submission.

These include the study protocol, patient recruitment procedures (such as advertising copy), the IB, résumés for Dr. Chase and her study staff, and the customized informed consent form (ICF). Sites are responsible for creating an ICF that describes the study treatment, potential study risks, and benefits. CanDo Pharma, like many sponsors, provides the sites with a study-specific, sample ICF that can be easily modified to meet the sites' requirements and IRB's preferences. Sally submits the customized ICF, along with the other relevant documents, to her local IRB.

The IRB members review those documents to ensure that the rights and safety of the patients are protected. If the IRB has questions or concerns, Sally gathers the appropriate information, prepares a formal response, and works with the IRB to resolve those issues. Dr. Chase and Sally may enroll patients only after the IRB approves the study.

In addition to IRB approval, CanDo Pharma must confirm that the clinical site staff has been properly trained and thoroughly prepared to conduct the study. CanDo Pharma schedules an investigator meeting for Dr. Chase and Sally, along with all the PIs and study coordinators from the participating sites in the United States, Australia, and Germany. This meeting lasts a full day and usually takes place at a nice hotel, centrally located between all the participating sites. Representatives from CanDo Pharma (primarily Dr. Abernathy and Nancy) explain each part of the study protocol in detail and answer the PIs' questions.

For part of the day, Sally attends a training session specifically for the study coordinators. They learn how to perform unusual study procedures, use study-specific laboratory kits, administer the experimental drug, and complete the CRFs. In some studies, such as this one, the study coordinators must enter the study data directly into a computer, rather than writing on paper case report forms. This type of system, called electronic data capture (EDC), requires additional training for Sally and her colleagues.

Patient recruitment is often included as an agenda item at investigator meetings. Whereas IRBs/IECs review recruitment strategies from a regulatory perspective, sponsors are concerned about the clinical sites' ability and commitment to enroll the number of patients needed to complete the study. Therefore, CanDo Pharma, asks the PIs to create and sign a patient recruitment plan. In the plan, the PI and study coordinator outline the activities they will use (such as advertising, website listings, and personal discussions) to ensure that potential study patients know about the study and how to volunteer.

Having attended the investigator meeting, Dr. Chase and Sally are now qualified to begin enrolling patients at their site. On other studies, the sponsoring company might arrange for individual site initiation visits, rather than the group investigator meeting. The site initiation visit has the same purpose as an investigator meeting, but a CRA conducts individual training at each PI's location. In either case, once the sponsor is certain that the site staff has been trained, the PI has signed the study contract, and the site's IRB has approved the study, the sponsor ships the experimental drug and placebo supplies to the site.

Patient Selection

After Sally receives the experimental treatment supplies, she begins lining up patients for the study. She has already identified a number of potential patients by combing through Dr. Chase's medical files. In addition, Dr. Chase has discussed the protocol with patients during their routine office visits, and several have expressed interest in the study. Sally contacts each of these patients, explains the purpose of the study, and schedules appointments for those willing to be evaluated further.

At the first patient visit, Sally gathers additional information to determine whether the patient meets the protocol's enrollment requirements. Some patients have medical conditions or other characteristics that disqualify them (called exclusion criteria). For example, the CanDo Pharma protocol requires Sally to exclude patients who are more than sixty-five years old. Patients must also have the desired disease characteristics (called inclusion criteria). For example, the protocol specifies that Sally may include only patients who have shown disease symptoms for more than two years. For patients who meet the inclusion and exclusion criteria, Sally explains the study in greater detail using the ICF.

The ICF explains the procedures in easy-to-understand language and the treatments that the patient will receive if he or she agrees to participate in the study. It describes the known effects of the drug, the length of the study, the potential risks and benefits of participating, and the number of required visits to Dr. Chase's office. One of those risks is that the patient may be assigned to the placebo group and, because of the blinding procedures, will not know the treatment assignment until after the study concludes. The ICF also guarantees that the patient can voluntarily withdraw from the study at any time.

Sally patiently and courteously answers all of the patient's questions about the study. Although she is honest and impartial in explaining the risks, Sally makes each patient feel special and conveys her gratitude for their time and interest in the study. Those who agree to participate in the study sign the ICF, and Sally schedules the patient's screening visit.

At the screening visit, Sally collects another set of information required by the protocol. She takes the patient's detailed medical history, including medications that he or she is taking, previous medical conditions, and the patient's vital signs (such as weight, blood pressure, and pulse). She draws a blood sample and sends it to the laboratory for analysis. In addition, Dr. Chase conducts a thorough physical examination. The laboratory results and examination may reveal new information that disqualifies the patient. Only those patients who still meet all of the protocol's inclusion criteria and do not have excluding conditions will be officially enrolled in the study.

Because many other clinical sites are also enrolling patients, Tom, Can-Do Pharma's biostatistician, has prepared a master schedule for assigning each new patient to one of the treatment groups in a random manner. Tom's randomization scheme was incorporated into an automated phone system called interactive voice response (IVRS). Sally contacts IVRS, which instructs her by telephone to give the enrolled patient an appropriately coded treatment (representing either the experimental drug or the placebo). (See Chapter 6 for details of randomizing procedures, IVRS, and statistical analysis.)

Patient Treatment and Evaluation

Sally organizes a specific treatment and visit schedule for each enrolled patient. She carefully explains restrictions and precautions that the patient must follow when he or she takes the experimental treatment. Because she does not know whether the patient received the experimental drug or placebo (Sally and the patient are both "blinded"), she counsels each patient the same way. Based on the protocol's requirements, she schedules the patient's remaining office visits and laboratory tests, making sure that the dates and times are convenient for the patient. Sally knows that these patients are doing her a favor by participating in the study, and she does her best to make the schedule convenient for them.

Sally and the office receptionist work together to remind the patients before each scheduled study visit, and they respond quickly if an enrolled patient calls unexpectedly. Sally and Dr. Chase are available 24/7 through an answering service. On long-term studies that span holiday seasons, patients sometimes experience conflicts with their appointments. Sally accommodates their preferences in rescheduling appointments but stays within the requirements of the protocol.

When patients arrive for scheduled visits, Sally makes sure they are comfortable and keeps their waiting time to a minimum. Patients enrolled in clinical studies at her site have a special waiting room, which is well stocked with magazines, internet access, and a television. Children whose mothers are study patients may use a special supervised play area stocked with toys.

Sally spends up to an hour with the patient on each scheduled visit. The procedures and information that she gathers vary from visit to visit, according to the protocol's schedule. On the CanDo Pharma study, she takes a blood sample, measures blood pressure, and asks the patient a series of questions about how they are feeling on every visit. On some visits she must schedule additional special procedures, such as x-rays, bone density measurements, and exercise stress tests at the local hospital.

Although the study procedures may be inconvenient and occasionally uncomfortable, the patients enjoy participating and look forward to their visits with Sally. She makes every patient feel special and works hard to make their experience on the study as pleasant as possible. Sally knows the importance of keeping the patients motivated to stay in the study until the scheduled completion. This can be especially challenging for studies that require treatment for several years.

If Sally was only responsible for one patient on one clinical study, her job would be simple. However, Dr. Chase has successfully enrolled a number of patients in the CanDo Pharma study, and they were each screened and enrolled on different days. Sally must keep track of the scheduled visits for all of these patients, doing the right procedures at the right times, based on their individual starting dates. In addition, Dr. Chase is the PI on several other ongoing studies, and Sally coordinates the visit schedules for those patients as well.

CanDo Pharma and the sponsors of Dr. Chase's other clinical studies have provided Sally with a number of tools to help her organize her work efficiently. For example, she has a study binder, which is her main reference book for the clinical study. It contains a complete set of study documents (e.g., the signed protocol, the final ICF, the investigator's brochure, a reference list of the normal ranges of laboratory values, study-related correspondence, a log of screened patients, and a log of the experimental drug inventory). It also contains the documents required by regulatory authorities such as Dr. Chase's résumé, the IRB approval letter, the laboratory's certification license, and a list of the IRB members. The CanDo Pharma team included labeled section dividers in its study binder so that Sally can add new information and documents as the study progresses.

In addition, Sally has prepared an individual folder for each patient. Each folder contains quick reference materials such as a summary of the protocol, a chart showing the study design, and the study's inclusion/exclusion criteria. Sally has also included blank laboratory requisition forms, preprinted specimen labels, and a set of worksheets to help her collect the required study data. On each scheduled visit, Sally takes the patient's folder and uses the forms, labels, and worksheets that apply for that patient visit.

Some sponsors provide Sally with paper CRFs; in those cases, she writes her data entries directly on the forms. However, the CanDo Pharma study uses electronic data entry (EDC). After the patient's visit, Sally logs onto CanDo Pharma's secure website and enters the information from her worksheets and the patient's chart into the electronic CRF pages. Unlike paper CRFs, the computer automatically checks Sally's entries for unreasonable values (for example, a positive pregnancy test in a male patient) and allows her to correct typographical errors immediately.

With EDC technology, CanDo Pharma collects the clinical results from all of the clinical sites worldwide more quickly and with fewer errors than paper CRFs. Sally and Dr. Chase like the computerized system, too. Their payments for conducting the study are based on the number of patients who have completed each office visit. The EDC entries immediately indicate which visits have been completed, and Nancy, the CanDo Pharma study manager, authorizes the corresponding payments.

Monitoring and Adverse Event Reporting

As another check of the study's progress, David makes scheduled visits (called monitoring visits) to the clinical site throughout the course of the CanDo Pharma study. (You can get a more detailed view of David's CRA job in Chapter 4.) David is responsible for confirming that the site is following the protocol and completing all documents accurately. Sally welcomes David's visits and uses this opportunity to discuss parts of the protocol that are unclear and procedures that have proven difficult to follow. David's suggestions and guidance allow Sally to make the appropriate adjustments while the study is ongoing. This keeps little problems from becoming more complex and possibly threatening the validity of the study.

Between David's visits, the data management group at CanDo Pharma reviews Sally's CRF entries. (For details about data management activities, see Chapter 5.) They send formal questions (called queries), asking Sally to explain missing information and to clarify data entries that do not comply with the protocol. For example, a blood sample may have been collected at an incorrect time, or an instrument may have malfunctioned on the day of the patient's visit. Sally answers many of the queries immediately via the secure website. For more complex queries, Sally and David discuss the issues during his monitoring visits before she responds.

Having a good relationship with the sponsor is especially important when unusual or unexpected things happen during a clinical study. If a patient experiences an unpleasant reaction, even if it is not caused by CanDo Pharma's drug, Sally and Dr. Chase must document and report this adverse event (AE). In these situations, David plays a critical role. Like Sally and Dr. Chase, David is available 24/7, directly or through his answering service. He is the link between the investigator site and all of the experts at CanDo Pharma regarding discussions and decisions about appropriate medical care for the patient. (For details about jobs dealing specifically with clinical safety, see Chapter 9.)

Sally's job is to document the AE. She makes the appropriate entries on the forms in the study binder and CRFs. She ensures that Dr. Chase completes and signs all documents in the patient's chart, including notes regarding special tests and treatments for the AE. Sally forwards to David and the CanDo Pharma physicians all the information they request regarding the patient's condition and treatment. Finally, she complies with requirements for reporting information about the AE to the local IRB.

In addition to regular monitoring visits by David, the investigator sites may also be visited by auditors from CanDo Pharma's clinical quality assurance (CQA) department or inspectors from regulatory agencies. Sometimes, these visits are prompted by a concern or complaint about the study or site. However, the auditors and inspectors also make scheduled visits to selected sites as part of their routine oversight, to confirm that the sites are complying with all regulatory requirements. (See Chapters 7 and 8 for details on CQA audits and regulatory inspections, respectively.)

When the last patient at Dr. Chase's site completes his or her last visit according to the protocol's schedule, Sally works with David to "close-out" the study at her site. Sally enters the remaining data through the secure EDC website, adds the final pages to her study binder, oversees shipment of the remaining blood samples, and schedules the close-out visit with David. She returns unused drug, blank forms, and extra laboratory supplies to CanDo Pharma. After David's close-out visit, Sally informs her local IRB that the study has closed and stores her study binder (which now has all of her patients' documents) for future reference. As a courtesy, Sally also sends a note to her study patients, thanking them for participating in the study.

ONE OF SALLY'S DAYS

Some study coordinators only spend part of their time conducting clinical studies, and the rest of the time they work as a regular staff member at their clinic or hospital. However, Sally works full-time for Dr. Chase as a study coordinator. Today, she plans to spend the day on several tasks: hosting a CRA who is monitoring one of Sponsor X's clinical studies, meeting with two study patients who have appointments for study assessments (one of whom is on the CanDo Pharma study), continuing work on a sponsor survey for a new study, preparing for next week's IRB meeting, and scheduling the first visit for a new patient on the CanDo Pharma study.

Sally arrives early at the office to review her study binder one more time before the CRA arrives. On the CRA's previous monitoring visit, she noted that the IRB membership list had not been updated and Dr. Chase had not signed the delegation of authority form for a newly hired physician who is assisting with the Sponsor X study. Sally makes sure those documents are in order and that all of the new documents in the binder have been properly dated and signed.

The CRA arrives on time, and Sally escorts her to a table in the staff's coffee break room. It's not an ideal room for the CRA to conduct her monitoring work because of the noise and heavy traffic, but space is limited in the office. Throughout the day, Sally will deliver and retrieve the documents that the CRA requests for her monitoring activities. Sally agrees to be available after lunch to escort the CRA to the pharmacy so that she can complete her drug accountability assessment.

After delivering the first batch of study documents to the CRA, Sally has a few minutes to call the new study patient. Yesterday, the patient's screening laboratory results came back, and Dr. Chase confirmed that the patient still met all of the enrollment criteria. Sally leaves a message on the patient's voice mail, explaining that he has been accepted into the CanDo Pharma study and suggesting several times for him to come for his first study visit.

The receptionist notifies Sally that her scheduled study patient has arrived. This patient has been treated with the CanDo Pharma drug for 6 months, so both she and Sally are familiar with the assessment procedures. Sally goes through her usual list of questions and measurements. However, before she sends the patient downstairs to the laboratory to take the routine blood sample, Sally notices a skin rash on the patient's arm. She knows that CanDo Pharma is concerned about skin rashes, and she informs Dr. Chase. After a careful examination, Dr. Chase confirms that CanDo Pharma needs to be notified about the rash. Sally asks the patient to wait while she telephones David, the CanDo Pharma CRA.

As Sally passes by the break room on her way to her desk, the Sponsor X CRA asks her for another batch of documents. At her desk, Sally listens to the reply from the new CanDo Pharma patient, who returned her call immediately and is excited about starting the study. He accepts the earliest date she proposed: tomorrow. Sally calls David with the good news (i.e., a new patient enrolling) and the bad news (i.e., the skin rash). She gives him a detailed description of the skin rash. David thinks that the rash is similar to the reports from other patients and will only require standard treatment. But he asks her to keep the patient in her office while he consults with Dr. Abernathy. He promises to get back to her quickly.

It is now approaching noon, and Sally takes a few minutes to book the new study patient into her schedule for tomorrow. She also calls the IVRS telephone number to report the new patient and obtain instructions for his randomized treatment assignment, which she logs in her study binder. Although Dr. Chase's patient appointments are kept in an electronic calendar, Sally stops by the front desk to alert the receptionist about tomorrow's appointment of the new study patient and asks the receptionist to call the patient to confirm the date and time. She drops off the additional documents to the Sponsor X CRA and updates Dr. Chase regarding David's instructions. She apologizes to the skin rash patient for the delay and takes her order for lunch while they wait for CanDo Pharma's instructions.

Not wanting to miss David's call, Sally eats a sandwich at her desk and works on the documents for the IRB meeting. Sponsor X recently updated the IB to include new data about the drug's adverse effects.

Because the new data may raise concerns about the drug's safety, Sponsor X wants this information communicated to all study personnel and their respective IRBs as soon as possible. In addition, Sally must revise the study's ICF to inform patients about the new side effects of the drug. Dr. Chase knows that the IRB will have questions about the patients that she has treated with the Sponsor X drug and has asked Sally to compile a summary of her patients' adverse events. In order for the IRB to review the AE information, the updated IB, and the revised ICF at next week's meeting, Sally needs to submit those documents today.

David calls to say that Dr. Chase can move forward with prescribing a standard antiseptic cream to treat the skin rash but asks her to follow the patient closely to make sure the condition does not worsen. Sally returns to the skin rash patient just as she is finishing her lunch, gives her the prescription from Dr. Chase, asks her to call immediately if the rash gets worse, and schedules an appointment next week to check on the progress of the skin treatment. She then releases the patient to go to the laboratory for her routine blood sample.

Fortunately, the patient on Sponsor Y's study is late, and the Sponsor X CRA is not yet finished with her document review in the break room. So, Sally has time to log onto the CanDo Pharma secure website and enter the data she collected this morning from the skin rash patient. In addition to the routine study data required for the six month patient visit, Sally adds comments on the adverse events pages of the EDC form regarding the skin rash observations. The website's mailbox also alerted her about queries that the CanDo Pharma data management staff posted regarding her earlier data entries. Most of the queries are straightforward, but she needs to track down information from the patient's medical chart to answer them properly. Answering the queries will have to wait until she has a break in her schedule.

As she is logging off, the receptionist stops by to say that one of the Sponsor X patients called to request rescheduling her next appointment and offered some alternative dates. Sally reviews the study protocol's scheduling requirements and gives the receptionist guidance on dates that fall within the protocol's specified range. The receptionist also mentions that the Sponsor Y patient has now arrived and reminds Sally that the Sponsor X CRA still needs to visit the pharmacy.

Because of the patient's late arrival and the laboratory's closing time, Sally asks the Sponsor Y patient to go to the laboratory first for his blood sample procedure and then return to the office for his other study assessments. In the meantime, Sally escorts the Sponsor X CRA to the pharmacy. Fortunately, the pharmacist is available to assist the CRA while she monitors the drug accountability records.

The Sponsor Y patient is waiting when Sally returns from the pharmacy. She apologizes for the delay and goes through the procedures for his scheduled visit. On this visit, he is due to receive a new supply of the blinded Sponsor Y treatment (drug or placebo). After she finishes her assessments and Dr. Chase reviews the collected data, Sally authorizes the pharmacist to issue the blinded supplies to the patient.

Sally and Dr. Chase are reviewing the IRB submission documents when the Sponsor X CRA returns to discuss the findings from her monitoring visit. They take advantage of this meeting to discuss the drug's newly observed adverse events and Sponsor X's perspective on their significance. The IRB may ask detailed questions, to ensure patient welfare, and Dr. Chase wants to be prepared to address the issues as thoroughly as possible. The CRA provides many helpful details.

Regarding her monitoring visit, the Sponsor X CRA compliments Sally for her excellent work in maintaining the site's study documents. She acknowledges that the issues cited during her last visit have been successfully addressed, and she found no new issues during the current visit. She also agrees to be available next week and respond quickly if the IRB requires more information about the Sponsor X drug's safety.

By the time the CRA leaves, Sally has just enough time to submit the documents to the IRB administrator for the IRB meeting. Sally could stay late to log onto the Sponsor Y secure website and enter the data from this afternoon's patient, but she decides to wait until tomorrow morning. Hopefully, she will have enough time to finish the entries before the new CanDo Pharma patient arrives. Also, she still has the sponsor survey on her desk for a new study that Dr. Chase wants to participate in. Sally has completed most of the easy items, but she still needs to go through the office records to fill out the most important information: Dr. Chase's patient enrollment performance on previous studies. She will try to find time to do that tomorrow.

SALLY'S BACKGROUND AND FUTURE

Unlike all of the other clinical positions described in this book, study coordinators are not industry employees. However, they play a vital role in the success of all clinical studies. Moreover, experience as a study coordinator can be a valuable stepping stone to help you qualify for entry-level clinical positions at drug, medical device, and CRO companies. As you will see from Sally's example, study coordinator positions are widely available, have fewer job prerequisites than industry jobs, and are a great way to quickly gain valuable clinical study experience.

Requirements

Sally has a bachelor's degree in biology and started working in Dr. Chase's office shortly after graduating. Most study coordinators are trained as a registered nurse, nurse practitioner, physician's assistant, medical technologist, or pharmacist. For those who have previous training as a healthcare professional and may also have a professional license, the transition to a study coordinator position is a natural progression, because they are already working in a place where clinical studies are often conducted.

In addition to an academic degree and the general qualifications discussed in Chapter 1 (see Box 1-2), study coordinators benefit from a basic knowledge of medical terms, clinical procedures, and clinical laboratory tests. However, Sally did not have previous clinical training, so she initially worked in Dr. Chase's office as a volunteer. How did Sally find Dr. Chase?

Finding a Study Coordinator Position

The best places to find study coordinator jobs are at institutions that conduct many clinical studies, such as university medical centers, hospitals and clinics in large metropolitan areas, and large medical practices. These clinical research centers employ large numbers of study coordinators, and they often have open positions. Or, you can ask your family physician. You may be surprised to learn that your physician is a PI or can refer you to medical colleagues who need a study coordinator.

ClinicalTrials.gov, a government-sponsored website, lists all clinical studies filed with the Food and Drug Administration (FDA), including the name and address of each participating PI. This database can be searched by location, and you can easily find PIs who are currently conducting clinical studies; many probably need study coordinators.

BOX 3-2. *Tips for Getting a Study Coordinator Position*

- Bachelor's degree in science or licensing as a registered nurse, nurse practitioner, physician's assistant, or pharmacist
- Experience interacting with patients is helpful but not required
- Consider volunteering in a doctor's office or clinic that conducts clinical studies
- Focus on entry-level jobs at large medical centers
- Aim to work under experienced study coordinators
- Take online courses on Good Clinical Practices
- Learn medical terminology
- Learn practical clinical study procedures such as phlebotomy
- Seek certification as study coordinator
- Ask physicians about ongoing clinical studies or referrals to colleagues who conduct clinical studies
- Search www.ClinicalTrials.gov for ongoing studies, active clinical sites, and PI addresses

Landing the Job

Look for an entry-level position, which will allow you to train under more experienced study coordinators. That's what Sally did. Dr. Chase's medical group maintains a staff of several study coordinators. As a volunteer, Sally initially worked under the supervision of Dr. Chase's senior study coordinator, who happened to be a registered nurse. Sally's first assignments were nonmedical tasks such as searching through medical charts for eligible patients, filling out sponsor surveys, setting up study files, filing patients' medical charts, and scheduling patient visits. These were valuable assignments, because they taught Sally about good clinical practice (GCP) requirements and the documentation skills needed for all clinical studies.

Because of Sally's enthusiasm and dependability, Dr. Chase soon offered her a study coordinator position, which became available when one of her coworkers left. Healthcare professionals who work at a hospital, clinic, or medical center often land their first study coordinator job by being in the right place at the right time. When staff physicians decide to conduct clinical research, either an independent or industry-sponsored clinical study, they are most likely to ask people they know and trust to assist them as study coordinators.

Sally supplemented her on-the-job training by studying the guidelines for GCPs and earning a phlebotomist certificate, which qualified her to draw blood samples. (Several websites offer online training for GCPs and clinical site skills.) She also earned a study coordinator certificate by taking courses sponsored by the Association of Clinical Research Professionals (ACRP). As Sally gained experience, her responsibilities grew, and she even began training some of the newer study coordinators. Eventually, she became the designated study coordinator on some of Dr. Chase's studies, including the one sponsored by CanDo Pharma.

Working for Industry

Sally now has 3 years of experience as a study coordinator and has supported a wide variety of studies in Dr. Chase's office. During that time, she has not only fostered a good working relationship with her coworkers and the physicians in Dr. Chase's group, but she has also worked closely with a number of CRAs, who monitor her site and work for sponsoring companies or CROs.

Some study coordinators enjoy having direct interactions with patients and prefer to continue working as a healthcare provider. Certainly, the success of all clinical studies depends heavily on the skills of high-quality study coordinators at investigator sites. However, Sally, like many experienced study coordinators, wants to become more involved with clinical studies and sees more opportunities for career advancement by working in industry. For her, the benefits (outlined in Chapter 1) of those positions are more attractive than working as a study coordinator.

CRAs provide a wealth of information about industrial job opportunities, and often they actively recruit study coordinators for clinical positions at their own companies. The CRAs have told Sally about several open positions, and as a result of her performance as a study coordinator, several hiring managers have asked her to consider CRA positions at their companies. Because of her study coordinator experience, Sally is well qualified for a sponsor or CRO-based position as a CRA. However, she would also qualify for other entry-level positions at those companies, such as regulatory affairs, data management, and clinical quality assurance. Study coordinators who have a nursing background also qualify for positions in clinical safety. These and other clinical study specialties are described in more detail in the following chapters.

STUDY COORDINATOR RESOURCES

Good Clinical Practice Training and Certificate Programs (Including CME Credits)

ClinfoSource (www.clinfosource.com) offers online GCP courses and certification for CME and CNE credit at reasonable cost.

Kriger Research Center International (www.krigerinternational.com) offers online GCP courses and certification in multiple languages on the country-specific web pages under the "Courses" tab.

Medical Research Management (www.cra-training.com) offers online and classroom GCP training for ACPE credit.

GCP Training Online (www.gcptraining.org.uk) offers online GCP training and certification, emphasizing the European Clinical Trial Directive, at reasonable cost.

Institute of Clinical Research (www.instituteofclinicalresearch.com) offers online CGP and European Clinical Trial Directive courses and certification at reasonable cost.

Study Coordinator Courses and Certificate Programs

Association of Clinical Research Professionals (www.acrpnet.org) offers online courses for CME, CNE, and ACRP credit and an industry-recognized clinical research coordinator certificate program at reasonable cost.

ClinfoSource (www.clinfosource.com) offers online site personnel courses for CME and CNE credit and a certificate program at reasonable cost.

Kriger Research Center International (www.krigerinternational.com) offers an online clinical research coordinator program and international diploma in multiple languages on the country-specific web pages under the "Courses" tab.

Skills Training for Site Coordinators

Phlebotomy Central (www.phlebotomy.com) provides a variety of educational materials on phlebotomy technique and information about reputable agencies that certify phlebotomists.

ClinfoSource (www.clinfosource.com) offers an online medical terminology course for CME and CNE credit at reasonable cost.

Kriger Research Center International (www.krigerinternational.com) offers an online course in medical terminology in multiple languages on the country-specific web pages under the "Courses" tab.

Free-ed.net (www.free-ed.net/free-ed/healthcare/medterm-v02.asp) offers an online course in medical terminology and CEU credit at reasonable cost.

Des Moines University (www.dmu.edu/medterms) offers a public access, online course in medical terminology.

Active Clinical Site Locations

U.S. National Institutes of Health/ClinicalTrials (www.clinicaltrials.gov) is a public-access, searchable registry of clinical studies and provides information about the study's purpose, locations, site contact information, and other details.

Salary Surveys for Study Coordinators

Applied Clinical Trials Salary Survey (www.appliedclinicaltrialsonline.com) is published annually. Access the most recent survey by entering "salary survey" in the website's Search field.

Salary.com (www.salary.com) publishes salary ranges for positions in a wide range of industries. Enter "clinical research coordinator" and if appropriate your targeted zip code in the Salary Wizard.

4

Entering as a Clinical
Research Associate

A CLINICAL RESEARCH ASSOCIATE (CRA) is the person responsible for verifying that the investigator sites conduct the study according to the protocol and ensuring the integrity of the clinical data at the sites. Because of the expense involved in conducting clinical studies and the need to protect patient welfare, sponsor companies like CanDo Pharma must ensure that the clinical sites are implementing the study protocol properly and providing sufficient safeguards during patient treatment. CRAs, whether they are employed by the sponsor company or a contract research organization (CRO), represent the sponsor when they monitor study activities at the clinical sites. Through those visits and ongoing communications, CRAs also serve as an important liaison between the site staff and the clinical study team. In addition to confirming proper study conduct, they highlight errors in tasks, documents, and data; track progress of patient enrollment and treatment; and ensure the sites take appropriate actions when problems or errors occur.

David is a CRA assigned to Nancy's clinical study. He works under the direction of Dr. Abernathy, the CanDo Pharma physician, and Nancy, the clinical study manager. David is employed by CanDo Pharma, but the company could hire temporary CRAs from a CRO to monitor clinical studies on its behalf.

THE ROLE OF THE CLINICAL RESEARCH ASSOCIATE

David's interactions with the other members of Nancy's study team are illustrated in Figure 4-1. His responsibilities begin during study start up and include evaluating potential study sites, collecting documents

BOX 4-1. *Clinical Research Associate Responsibilities*

- Serve as the communication link between sponsor and PI
- Train site staff on protocol procedures and requirements
- Ensure clinical sites receive study materials and drug (or medical device) supplies
- Collect regulatory documents
- Monitor clinical sites to confirm:
  - PI and staff qualifications
  - Protocol compliance
  - Patient safety
  - Compliance with patient inclusion/exclusion criteria
  - Informed consent form signatures and dates
  - Source document verification
  - Case report form accuracy and completeness
  - Drug (or medical device) accountability
  - Adverse event reporting
- Report and resolve protocol deviations
- Prepare monitoring reports
- Maintain master study file
- Track and report status of enrolled patients
- Conduct site close-out visits
- Arrange for return of experimental drug or medical devices
- Archive study documents

required by regulatory authorities, advising the designers of the case report forms (CRFs), and participating in the investigator meeting. He helps the principal investigators (PIs) with their arrangements for the study by providing study supplies and facilitating shipment of the drug and placebo treatments. During the study, he periodically visits the sites to check the study documents, drug storage, and collected data. David also takes advantage of these visits to discuss any problems the PIs may have encountered. After the study, he officially closes the sites by confirming that all data are collected, all unused materials are returned, and all study documents are archived.

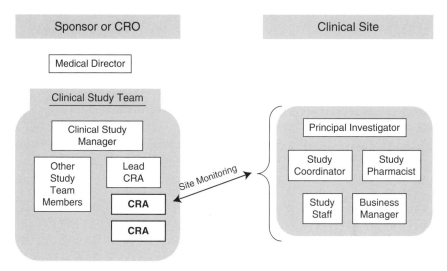

Figure 4-1. CRA interactions with other members of the study team.

Because Nancy's study includes sites in three countries, David can rely on a CanDo Pharma CRA based in Germany and another in Australia to monitor the clinical sites in those countries. Although he has no supervisory responsibilities, he coordinates his activities with the other two CRAs. The clinical department also provides a staff of study support specialists who assist the CRAs with clerical and administrative tasks.

The costs of Nancy's clinical study are budgeted individually by each work unit in CanDo Pharma's clinical department. David's purchasing authority is limited to small office items, but he often receives authorization from Nancy or Dr. Abernathy for more expensive purchases of essential study materials and equipment. During his frequent monitoring visits, David must follow CanDo Pharma's travel expense guidelines, but CanDo Pharma reimburses him for all reasonable travel expenses. Because travel is essential for his job, the clinical department incorporates those costs in its annual operating budget.

Site Evaluation Visits

By the time David joins Nancy's study team, Dr. Abernathy has approved the final study protocol and his first task is to conduct site evaluation visits. Only qualified sites may participate in the clinical study, so sponsors evaluate all clinical sites for their suitability. Dr. Abernathy has suggested some potential sites, and David first consults the lists published by regulatory agencies to ensure that none of those PIs have been disqualified or restricted due to previous clinical study misconduct. For PIs he deems viable, David schedules site evaluation visits after the respective investigators have completed CanDo Pharma's survey and indicated they are interested in participating.

During his visit, David confirms that the PI and the key site staff have the appropriate credentials and that they can devote sufficient time to the study. He reviews the résumés of the PI and, if applicable, coinvestigators who will participate in the study. They not only must have valid healthcare licenses but also clinical qualifications that are appropriate for the clinical study. A board-certified cardiologist, for example, may not be suitable as the PI on a cancer study. If the clinician is already serving as the PI on several other studies, either for CanDo Pharma or other sponsors, David asks the PI to explain how the site staff will manage their time between these studies and the measures they will take to ensure the confidentiality of each sponsors' study files. Finally, David checks to make sure the PI has completed the appropriate financial disclosure forms, indicating whether the PI owns stock or has other financial interests in CanDo Pharma's business. Personal financial investments are permitted; however, because this might affect the PI's impartiality, CanDo Pharma must decide whether to include such investigators in its study and, if so, take precautions to ensure the integrity of the study is not compromised.

David also determines whether the site's facilities are suitable by inspecting the areas where patients will be treated, the laboratory samples will be processed, and the experimental drug will be stored. For example, Nancy's study requires bone density measurements and an exercise stress test, so David confirms that the site is properly equipped for these procedures. Because the collected blood samples must be centrifuged and refrigerated immediately, David confirms that the proper laboratory facilities are nearby. He also checks to make sure the pharmacy has the appropriate environmental and security controls and that the drug cabinets are locked.

Finally, he reviews the PI's strategy for recruiting patients, to confirm that the site will be able to find eligible patients. Some sites advertise in the local media; others contact civic and patient advocacy groups in the community to generate interest in a new study. David compares the recruiting strategies the site used on previous studies with the number of qualified patients it enrolled. Every study is different, but a site with a successful history of finding eligible patients is more likely to enroll patients in a new study, too.

After each site evaluation visit, David submits a written report of his findings to the team. The decision whether to select the site or not is made by Nancy and, ultimately, Dr. Abernathy. But David's recommendation is an important factor in that decision.

For each selected site, David works with the site's study coordinator to collect the documents required by regulatory agencies. These include résumés from the PI and other key site personnel, a copy of the protocol signed by the PI, the local institutional review board (IRB) approval letter, a blank informed consent form approved by the IRB, the laboratory's certification license, and the normal ranges of laboratory results. Some regulatory documents are stored in CanDo Pharma's files; others are stored in the PI's study file. Before the study starts, David must confirm that all required documents have been collected and properly filed.

In parallel with his activities at the clinical sites, David offers suggestions to Maria, the team's data manager (see Chapter 5), while she is designing the CRFs. Because the team has decided to use electronic CRFs for Nancy's study, the site coordinators will enter their collected data using a computer terminal. David's perspective is especially valuable to the CRF designers because he understands the work environment at the investigator sites. His suggestions ensure that the data entry procedures are easy for the coordinators to follow.

Planning the Investigator Meeting

After the study team has engaged all of the clinical sites, it hosts an investigator meeting for the PIs and clinical site study coordinators. (See Chapter 3 for details of the PI and study coordinator positions.) Dr. Abernathy and Nancy plan the agenda and make the arrangements for the meeting, respectively, but David is a key participant.

At the meeting, David leads the breakout sessions with the study coordinators. He reviews all of the study's treatment and drug handling procedures. The coordinators receive specific training on specialized medical procedures and how to store, dispense, and administer the drug (and placebo) treatments. Because some of the patients' blood samples will be sent to a central laboratory for special analysis, David explains the sample preparation and shipping procedures in great detail. He has also made arrangements to give the study coordinators hands-on computer training on how to enter the study data using electronic data capture (EDC).

Study Materials and Master File Preparation

In parallel with the investigator meeting, David prepares a variety of study materials, some of which are used by the investigator sites and others that are used by the CanDo Pharma team. For each site, he prepares a study binder that includes all of the reference materials needed by the study coordinator. These include the protocol signed by the PI, the final informed consent form (ICF), and the investigator's brochure. The study binder contains tabbed sections for the study coordinator to insert the PI's résumé, the IRB approval letter, the laboratory's certification license, a reference list of the normal ranges of laboratory values, and a list of the IRB members. He also includes tabs for information that will be generated during the study (such as study-related correspondence, a log of screened patients, and a log of the experimental drug inventory).

David orders custom-designed kits for collecting the special blood samples and preprinted shipping labels to identify each specimen. Some sites do not have the computer equipment needed to access CanDo Pharma's electronic CRFs. For those sites, David makes arrangements to install the appropriate EDC hardware and software. All of these materials must reach the investigator sites before starting the study.

At CanDo Pharma, David sets up the study master file. This file contains all of the official documents associated with Nancy's study. CanDo Pharma's archive department has defined a specific and detailed classification system for master study files. The standardized archive structure simplifies filing and retrieving individual documents during and after the study. Although many team members will submit documents to the master file and Nancy is ultimately responsible for its content, David is the primary person who maintains the master study file during the study and makes sure all new documents are filed in a timely manner.

Regulatory Requirements Compliance

Nancy's team wants to start the study as soon as possible, but CanDo Pharma must comply with regulatory requirements, verifying that each site is properly qualified and trained before it enrolls and treats any patients. CanDo Pharma controls the clinical site's ability to start treating patients by controlling drug shipments. David completes a checklist, which confirms that the site has provided all of the required regulatory documents (i.e., protocol signed by the PI, IRB approval letter, signed site contract, and site evaluation documents) and that the site staff has been properly trained (i.e., at the investigator meeting or site initiation visit). Nancy, the study manager, and Amy, the regulatory affairs specialist (see Chapter 8), review the checklist for completeness and sign the authorization form to ship the clinical study treatments. Their authorization allows CanDo Pharma's manufacturing unit to prepare, package, and ship the coded supplies to the investigator site.

David follows this process closely—from completing the checklist to shipping the supplies—and intervenes if necessary to make sure that the sites receive their shipments as quickly as possible. He keeps accurate records of all drug and placebo shipments in the study master file. He also monitors the rate at which the sites use their supplies and facilitates additional shipments before their treatment supplies become depleted.

Regulatory agencies require sponsors to visit the investigator sites and monitor site activities during the study. Nancy has prepared a monitoring plan that explains the specific monitoring tasks the CRAs must accomplish during each site visit, consistent with Good Clinical Practices (GCPs) and CanDo Pharma's standard operating procedures (SOPs). By defining the protocol-specific monitoring activities, the monitoring plan ensures consistent study conduct between sites in all three countries. For example, the monitoring plan requires David to schedule monitoring visits to each site every 4–6 weeks. He must verify 100% of the data in the CRFs against original source documents and inspect all signed ICFs. The monitoring plan also sets requirements for reviewing the information in the study binder, open data queries, serious adverse event (SAE) reports, and the drug accountability records.

Source Document Verification

One of David's most important monitoring tasks is called source document verification. He compares the study data that the coordinator entered in the electronic CRFs with the original patient records (e.g., the medical chart, laboratory test results, and printouts from medical instruments), to verify that the data match. In addition, as David reviews the patient records, he looks for trends in the data and medical notes to confirm that the patient's safety is being appropriately assessed and followed up by the PI. If patients have missed a scheduled visit or did not complete a study procedure, the CRFs must include an explanation. He also checks to see that all medical conditions written in the patient's chart (e.g., emergency dental surgery or unscheduled pain medication) are recorded in the CRFs. David confirms that all the enrolled patients meet the protocol's inclusion/exclusion criteria and that they signed the ICF before participating in any study-specific procedures. If any patients have withdrawn from the study, he makes sure that an explanation is noted in the CRF.

Because CanDo Pharma's study uses electronic CRFs, David does his source document verification by logging on to the database while he is at the investigator site. When he has finished his review, he can electronically mark each CRF page that he verified, including those where he found errors. (For studies that use paper CRFs, David would verify the data on the completed CRF pages and bring them back to CanDo Pharma. Data management staff would then enter the data from the paper CRFs into CanDo Pharma's clinical database.) David also uses the completed CRFs as proof of the completed patient visits, so that Nancy can authorize paying the PIs for the work they have completed.

David inspects the documents in the site's study binder for completeness. He especially notes any changes since his last monitoring visit, such as recent notifications to the local IRB, changes in the PI's credentials, or new delegations of authority when the PI adds a new member to his or her staff. For changes such as these, the study coordinator must include properly dated new or revised documents in the study binder. David also inspects the documentation for handling the patients' blood samples. For Nancy's study, the sample processing logs, shipping records, and laboratory reports must all be included in the site's study files.

Drug Accountability

David's inspection of the site's drug handling procedures is called drug accountability. David visits the site's pharmacy to check the drug storage and security logs. He checks the inventory records to verify that the pharmacist is preparing and dispensing the treatments (drug and placebo) according to the study protocol. He also checks the expiration date on the treatment supplies and confirms that they are stored (e.g., refrigerated or protected from light) as stipulated in the protocol. The pharmacist must account for disposition of all of CanDo Pharma's drug and placebo supplies, whether they are in storage, consumed by the patients, destroyed, or returned to the sponsor.

Adverse Events and Problems

Sometimes patients experience unpleasant reactions (called adverse events, AEs) during treatment in a clinical study. David is especially attentive to these situations. During his monitoring visits, he verifies that the PI recorded all AE observations properly in the patient's chart and on the corresponding CRF pages. Some adverse events may be serious (called SAEs), requiring immediate medical treatment and rapid reporting to regulatory agencies. When an SAE occurs, whether David is at the site or not, he works with the PI to make sure everyone responds quickly to protect the patient's safety and file the appropriate regulatory reports. For all AEs and SAEs, David confirms that the study coordinator has made the appropriate entries in the study documents. (For further details about clinical safety activities, see Chapter 9.)

At the end of each monitoring visit, David reviews his findings with the PI and study coordinator. He always compliments them for all the study activities they are performing correctly. When he spots errors, David encourages the site staff to make the appropriate corrections, which they usually make easily and quickly; International Conference on Harmonization (ICH) regulations prevent him from making any changes himself.

David especially draws on his good relationship with the site staff to resolve controversial problems tactfully but firmly. If the PI disagrees or does not understand the need for changes, David may invite other CanDo Pharma team members to assist in resolving the issue. For each open issue, he must monitor the progress of discussions between the site and the CanDo Pharma team until it is resolved. And, he never misses an opportunity to express his gratitude to the site staff for their efforts.

After each monitoring visit, David submits a monitoring report to Nancy for her signature and files it in the study master file. The monitoring report not only documents his activities during the visit, but it also lists his key findings and the corrections that the PI made. He continues to list the status of all open issues in subsequent monitoring reports until they are resolved.

Most sponsors, including CanDo Pharma, watch patient enrollment closely. If the sites cannot find the number of patients required by the protocol, CanDo Pharma may be forced to cancel the study—an expensive and highly undesirable situation. As part of David's ongoing communication with the study sites, he tracks the number of enrolled patients and works with the study coordinators to identify factors that may be limiting patient enrollment. At the study team meetings, David reports the status of patient enrollment. If enrollment lags, the team relies on David's experience and diplomatic skills to work with the sites and find creative ways of locating additional qualified patients.

Study Site Close Out and Archiving Study Documents

After each clinical site completes the specified treatments and procedures on its last patient, David (under authorization from the study team) works with the study coordinator to close-out the study site. He confirms that the site staff has completed all the CRF entries and resolved all queries. Next, he gives the study coordinator specific instructions for returning unused study materials, forms, and drug supplies. David's final monitoring visit is called the close-out visit. He inspects the site to confirm that all of CanDo Pharma's documents and unused supplies have been returned and that the site has properly stored all the patient-related study documents. He records these activities in his close-out report, which he submits to Nancy for signature and stores in the study master file.

David's final responsibility on the CanDo Pharma study is to archive the study documents. In conjunction with Nancy, the study manager, he reviews all the documents in the study master file, collects and files missing items, and ensures that the file meets CanDo Pharma's archiving standards. The study master file becomes part of CanDo Pharma's official archive and must be available to auditors from regulatory agencies and the company's clinical quality assurance (CQA) department. (See Chapters 7 and 8 for further information about CQA and regulatory affairs activities, respectively.) When the final study summary becomes available, David sends it to the PIs for their files and their local IRB.

ONE OF DAVID'S DAYS

David spends about half of his time travelling to investigator sites to monitor Nancy's study. On those days, his work day is dictated by the site's normal office hours and the tasks detailed in the monitoring plan. Today, however, David is at his CanDo Pharma desk, and he plans to spend the day on two tasks: preparing his presentation for tomorrow's clinical study team meeting and scheduling his routine monitoring visits for the next few weeks.

Because the study team knows that the patient enrollment rate is critical for the study to stay on schedule, Nancy includes a standing agenda item on enrollment status, and David makes the presentation at each team meeting. Emails have arrived overnight from the CRAs in Germany and Australia, providing him with the current enrollment status at the sites in their regions. He just needs to combine that information with the details from the sites that he monitors and then prepare the slides for his presentation. The team wants to know how many patients have started treatment, how many have finished their treatment schedule, how many have dropped out of the study, which sites have enrolled the most patients, and which sites have not enrolled any patients.

Nancy has added another item to this week's agenda. A number of the study patients have developed urinary tract infections. The infections seem to respond well to standard antibiotic treatment, but the study team has two concerns. First, no one predicted that the drug would make patients more susceptible to infections; nothing in the accumulated data from earlier studies had suggested that the drug affected the patient's immune system. Second, although the infection itself is not posing a health risk to the patients, it might be an early sign of a more serious problem with the drug.

Nancy asked David to summarize these cases and present his findings to the study team. She wants to know how many cases have occurred, when the infection appeared in each patient with respect to treatment initiation, whether the infection recurred in any patients, whether the infected patients had a history of urinary tract infections or any other infections before entering the study, the demographics of the infected patients (i.e., age, gender, ethnicity), and which clinical sites have reported cases and which have not.

David spends most of the morning compiling the information he received from the other CRAs and gathered from his sites. The German cases are well documented. He has a few questions about the Australian cases, but the issues are too complex to explain in an e-mail to the Australian CRA. However, David is confident that a short telephone conversation will easily clarify those issues. Unfortunately, because of the time zone difference, he will not be able to call her until the end of the day. With noon approaching, he completes all the slides except those that require input from Australia.

Just as David is thinking about lunch, he receives an urgent call from Sally, the study coordinator at Dr. Chase's site. The study protocol requires measuring the patient's cardiac performance during an exercise stress test. To minimize variations in the data collection, Nancy's team requires the clinical sites to use a specific, state-of-the-art cardiovascular monitoring instrument. Sally has been having difficulties calibrating the equipment, and none of the procedures she learned during training at the investigator meeting have solved the problem. A patient will arrive for his scheduled exercise stress test on Monday, and Sally knows the cardiac instrument must be properly calibrated before then.

Several other sites have had difficulties with the cardiac instrument, and David suspects that Sally is experiencing the same problem. He gives her instructions that fixed the problem at the other sites, thinking that those instructions will be sufficient. Sally, after all, is one of the best study coordinators on the study, and he is confident she will follow through successfully. However, in a series of phone calls over the next few hours, it is clear to David that the problem Sally is experiencing is something different. With no other alternatives remaining, David finally decides he will need to troubleshoot the instrument personally.

His next monitoring visit to Dr. Chase's site is due in several weeks, and he proposes moving it up. (Although the monitoring plan sets an outer limit on the interval between monitoring visits, CRAs are free to visit clinical sites more frequently, as necessary.) He will plan to address the instrument problems and conduct his monitoring activities on the same trip. However, because his monitoring tasks take several days, David will need to be at the site over the weekend, when CanDo Pharma's and Dr. Chase's offices are normally closed. Nancy confirms that she will be available over the weekend by phone and e-mail to assist him, and Sally agrees to give him access to Dr. Chase's facilities and escort him after normal business hours. David also receives authorization from Dr. Abernathy to bring a new cardiac instrument with him, in case they need to replace Sally's instrument altogether.

David then rushes to make his travel arrangements and plans to leave tomorrow, immediately after Nancy's study team meeting. On his way to the CanDo Pharma storage facility to present his authorization papers and pick up the new cardiac instrument, one of CanDo Pharma's newly hired CRAs stops him in the hallway. He is having difficulty understanding the clinical department's SOP describing site evaluation visits. David takes a few minutes to explain several unfamiliar terms and answer the CRA's questions. He also tells the CRA where to find the referenced forms and templates in the department's electronic files.

When David returns to his desk with the boxed cardiac instrument, he calls the Australian CRA to ask about several of the urinary tract infection cases. A short time later, he finally has all the information he needs to complete the slides for his presentation. Nancy wants to discuss the infection information with him tomorrow before the team meeting, so he forwards his slides to her as an electronic file before he leaves.

Hopefully, he will also have time in the morning to schedule his next regular monitoring visits. To meet the 6-week requirement specified in his monitoring plan, he must visit three of his sites before the end of the month, but this scheduling can be tricky. He coordinates his schedule with the respective study coordinators and PIs to ensure that they will be available to meet with him on the days of his visit. Sometimes, David must do considerable negotiating to make his travel schedule efficient for him and also convenient for the respective sites. Also, in advance of each monitoring visit, he prepares and sends the sites an outline of the monitoring tasks he will be conducting, so that they can properly prepare for his visit. If he cannot settle these arrangements tomorrow morning, he will have to postpone scheduling those monitoring visits until next week, when he returns from his visit to Dr. Chase's site.

HOW DAVID GOT HIS CRA JOB

Requirements

After service as an army medic, David became a registered nurse. Many CRAs have a nursing background, but this is more a coincidence than a requirement; the places where clinical studies are conducted are also the

BOX 4-2. *Requirements for a CRA Position*

- At a minimum, bachelor's degree in science or a license in nursing, pharmacy, medical technology, or physician's assistant
- Clinical experience: 1 to 3 years of work experience in life science or a medically related field (e.g., study coordinator)
- Ability to understand technical, scientific, and medical information
- Familiarity with GCPs and ICH regulatory requirements
- Computer operation skill
- Ability to write and present scientific and clinical issues clearly using appropriate terminology
- Ability to deal with time constraints, incomplete information, and unexpected events
- Good organizational and planning skills
- Ability to work effectively in a team or matrix environment
- Experience interacting with patients is helpful but not required

places where many nurses work. If you have a bachelor's degree or training as a physician's assistant, pharmacist, or medical technologist, you can also become a CRA.

Shortly after he started his nursing career, David was asked to help with a new clinical study at his hospital. He liked working on clinical studies and eventually became the study coordinator for several studies at the hospital. A different CRA monitored each of the studies; some CRAs worked for a sponsoring company, and others worked at a CRO that had been contracted by a sponsor.

All of the CRAs enjoyed working with David and were impressed with his work. They valued his understanding of GCPs, accurate record-keeping, and eye for details. He carefully followed the study protocol and their instructions, but he was not shy about asking questions. He also offered thoughtful solutions to study problems such as poor patient enrollment. The monitors encouraged David to consider becoming a CRA and alerted him to job openings at their companies.

Through these interactions, David realized that becoming a CRA would be a good career move for him. As a CRA, he would be able to work with many different clinical sites, not just at one site. He liked the idea of traveling to different cities, visiting top medical centers, and interacting with leading medical experts. He also realized that CRA jobs paid well, had good employment benefits, and would give him experience that could lead to greater career opportunities within the clinical department. However, he knew that he would need to move to the city where the hiring company was located.

Finding a CRA Position

You can find entry-level CRA positions at large sponsor companies (bio-pharmaceutical and medical device companies) and at large CROs. Small companies always prefer CRAs with many years of experience. Most companies post job openings on their websites, which you can search by job title. The Pharmaceutical Research and Manufacturers of America (PhRMA), the Biotechnology Industry Organization (BIO), Medical Design & Manufacturing (MD&M), and the Association of Clinical Research Organizations (ACRO) are trade organizations for biopharmaceutical, medical device, and CRO companies, respectively, and list their member companies on their websites. (Chapter 13 lists the websites for PhRMA, ACRO, BIO, MD&M and their member companies.)

CROs offer several advantages over sponsor companies for entry-level CRAs. First, CRAs at a CRO have greater opportunities to work on many different types of study protocols in many different disease areas for many different sponsors. Second, CRAs get lots of monitoring experience in a short period of time, because CROs expect their monitors to spend 80% of their time visiting sites. Finally, CROs provide their workers with structured, systematic training, and they require CRAs to achieve specific skill standards before permitting them to monitor sites independently.

The hiring managers at both CROs and sponsor companies look for the same skills and qualities in their job candidates. Previous clinical study experience is highly desirable, even for entry-level CRA positions. Being a study coordinator, like David, qualifies. But, there are other ways. One approach is to work in a study support role. Sponsors, CROs, and even some large medical centers hire study support people to handle the extensive documentation activities involved with clinical studies. Although this is largely clerical work (and does not require a college degree), you gain practical experience and learn about clinical studies. (If you work in a study support position at a sponsor company or CRO, you may hear about new, entry-level CRA positions before they are officially posted.)

BOX 4-3. *Tips for Getting a CRA Position*

- Study coordinator experience is a big plus
- Consider volunteering in a doctor's office or clinic that conducts clinical studies
- Consider working in the laboratory of a physician who conducts independent clinical research (often located at medical schools and major medical centers)
- Consider a clinical study support (clerical-level) position in a clinical department
- Consider an IRB administrative support position at a hospital or medical center
- Take online courses on Good Clinical Practices
- Learn medical and clinical research terminology
- Seek certification as a CRA
- Join a CRA professional society such as SoCRA or ACRP
- Network with CRAs for job referrals
- Search CRA job postings on websites of drug, medical device, and CRO companies
- Preferentially look for entry-level jobs at large CROs and biopharmaceutical companies

---

Another way to gain relevant clinical experience is to work in a clinical research laboratory or to provide administrative support for a local IRB. Physicians at large medical centers sometimes conduct independent clinical studies and require assistance with processing laboratory samples. This is a good way to learn about clinical studies, and the clinicians may allow you to accompany them when they treat and assess their study patients. Alternatively, as an IRB administrator, you would handle all of the official clinical study documents on behalf of the IRB's chairman. It's a good way to learn about GCPs, study protocols, informed consent issues, and other regulatory procedures.

Landing the Job

Hiring managers know that entry-level CRA candidates do not have previous monitoring experience, even if they have experience with clinical studies. Therefore, they look for work habits, accomplishments, and other skills that are characteristic of a good clinical monitor.

All of the hiring managers ranked David as a highly desirable CRA candidate. Although he had no previous monitoring experience, he had solid clinical study experience as a study coordinator. He understood GCPs, knew how to follow a protocol, and worked effectively with his local IRB. Furthermore, the CRAs who monitored his clinical site confirmed that David kept accurate and complete records, submitted high-quality CRFs, and kept both the experimental drug and the patients' records secure. Finally, he had always shown good judgment—never ducking his responsibilities but always consulting and informing others when appropriate.

The hiring managers were less impressed with the candidates who competed with David for entry-level positions, even though some were also study coordinators. CRAs spend much of their time unsupervised and away from the office. Hiring managers, therefore, discarded candidates whose previous supervisors said they were untrustworthy, could not work independently, or could not manage their time wisely. CRAs must have excellent interpersonal skills, even in stressful situations. Therefore, hiring managers also eliminated candidates who were disrespectful, unfriendly, overbearing, or timid. Finally, CRAs must be excellent communicators. The hiring managers rejected candidates who submitted résumés with grammatical and spelling errors and who did not speak clearly and professionally during their interviews.

Working for a Sponsor Company Versus a CRO

Although David could have taken a CRA position at a sponsor company, he decided to work for a large CRO. Monitoring experience is essential for CRAs, and David needed to develop his monitoring skills from scratch. His first position at the CRO was called an in-house CRA. He worked in the CRO office, assisting CRAs who monitored clinical sites. He maintained the study master files, prepared study binders for the sites, tracked the status of IRB approvals and other regulatory documents, and sent status reports to the sponsor companies.

As part of his on-the-job training, David accompanied the senior CRAs on their monitoring visits. They assigned his monitoring duties, checked his work, and evaluated his performance. By watching and working under different CRAs, David not only learned monitoring skills but also observed a range of monitoring styles. He soon progressed from an in-house CRA to a CRA qualified to monitor independently.

During this time, David joined a CRA professional society. The Association of Clinical Research Professionals (ACRP) and the Society of Clinical Research Associates (SoCRA) are two prominent societies for CRAs. The advantages of society membership include training classes, webinars, an official CRA certification program, newsletters and other CRA-oriented publications, and an annual society meeting.

The two most effective ways for David to advance his career as a CRA are to earn CRA certification and maintain his listing in his professional CRA society directory. Certification is an additional qualification that David can add to his résumé. It gives him an advantage over other candidates if he decides to apply for CRA positions at other companies. Both ACRP and SoCRA offer CRA certification programs for their members. David's listing in his professional society's membership directory also makes him visible to recruiters and hiring managers who are looking for experienced CRAs. If he chooses to stay at the CRO, his career advancement will depend mostly on his on-the-job performance, taking advantage of the company's internal training programs, and performing well on critical tasks such as solving difficult problems and meeting sponsors' deadlines.

Although David performed well and could have advanced his career at the CRO, he saw some limitations. As a contract worker, David jumped from one study to another and from one sponsor to another. He wanted to follow one experimental product through its development to see it become a marketed product. Although the salary and benefits are good at CROs, the compensation packages at sponsoring companies are often better. Also, the CRAs at sponsor companies typically spend less time travelling than CRAs at most CROs.

Hiring managers at sponsor companies often prefer to hire experienced CRAs, because they want their new employees to begin monitoring immediately. They draw heavily on CRAs trained at CROs. Because David was listed in the membership directory of his CRA society, the CanDo Pharma hiring manager easily identified him as a potential candidate.

It's always easier getting a job when the job comes to you. Once you have clinical experience and monitoring experience as a CRA, your next CRA job is even easier to get. For CRAs with only a few years of experience, the choices for CRA positions are plentiful, varied, and quite attractive. Professional recruiting firms aggressively recruit experienced CRAs for their biopharmaceutical and medical device clients. You can likely find a location, employer, and job description that match your preferences. Why? The demand for experienced CRAs has always been greater than the number of CRAs available—a nice place to be!

MOVING FORWARD

David, like many CRAs, chose to enhance his monitoring experience and advanced to more responsible positions within the CRA track. He liked the opportunity to travel to top-notch medical centers and interact with well-known clinical investigators. As a senior-level CRA, he would have more opportunities to travel internationally and meet world-class investigators. Also, because of his accumulated experience in successfully solving problems on clinical studies in different therapeutic areas and different types of experimental products, David would be actively involved in team discussions for designing new studies, selecting investigators, and deciding how to proceed when presented with unexpected results. David would also have more responsibility and perhaps supervise other CRAs.

Companies differ in the titles for CRA levels (e.g., CRA-I, CRA-II, CRA-III, or Assistant CRA, CRA, Senior CRA), but promotion from one level to the next can occur every 2–3 years, based on performance. At the top of this ladder, CRAs usually assume responsibilities as a "lead CRA." A lead CRA monitors some clinical sites, but more importantly he or she also oversees and coordinates all of the other CRAs assigned to a large, multinational clinical study.

Some experienced CRAs are regional CRAs. They work from an office in their home and only monitor clinical sites within easy travelling distance from their home. Their employer (either a sponsor company or CRO) provides and maintains all of the equipment for their home office. Sometimes, the position includes use of a company car. Regional CRAs typically have greater flexibility in their working hours than those who work in a company office.

Other experienced CRAs choose to work as an independent CRA, either alone or for an employment agency. Like regional CRAs, independent CRAs also work from an office in their home, but they typically must set up and maintain their own office equipment. Besides the convenience of working from home, independent CRAs have the freedom to accept or reject assignments, allowing them to work either full- or part-time.

From the top of the CRA ladder, David can move into a management position. For example, some companies consolidate their regional CRAs in a field-monitoring department, supervised by a CRA manager. CRA department managers assign work, assess performance, and oversee supplemental training of their regional CRA staff. Alternatively, if David wants to take more responsibility for running clinical studies, he can consider a clinical manager position similar to the one that Nancy holds at CanDo Pharma, or a position in project management, clinical training, or quality standards. (See Chapter 11 for information about these positions.)

With his CRA experience, David could also advance his career in many other directions, some of which have greater potential for increased salary and benefits. He would have more opportunities to travel to scientific and medical conferences but would not be routinely travelling to investigator sites. He is well qualified to take a position in data management, regulatory affairs, or clinical quality assurance (see Chapters 5, 8, and 7, respectively, for information about these positions). If he is interested and a talented writer, David can consider a position in medical writing (see Chapter 10). Finally, because David has a nursing background, he might also consider a job in clinical safety (see Chapter 9). Any of these clinical career tracks would broaden his clinical and product development expertise, making him eligible eventually for high-level management positions, if he wishes.

CRA RESOURCES

Good Clinical Practice Training and Certificate Programs

ClinfoSource (www.clinfosource.com) offers online GCP courses and certification for CME and CNE credit at reasonable cost.

Kriger Research Center International (www.krigerinternational.com) offers online GCP courses and certification in multiple languages on the country-specific web pages under the "Courses" tab.

Medical Research Management (www.cra-training.com) offers online and classroom GCP training for ACPE credit.

Association of Clinical Research Professionals (www.acrpnet.org) offers online GCP courses and certification for CME, CNE, and ACRP credit at reasonable cost.

GCP Training Online (www.gcptraining.org.uk) offers online GCP training and certification, emphasizing the European Clinical Trial Directive, at reasonable cost.

Institute of Clinical Research (www.instituteofclinicalresearch.com) offers online GCP and European Clinical Trial Directive courses and certification at reasonable cost.

Medical Dictionaries and Terminology

ClinfoSource (www.clinfosource.com) offers an online medical terminology course for CME and CNE credit at reasonable cost.

Kriger Research Center International (www.krigerinternational.com) offers an online course in medical terminology in multiple languages on the country-specific web pages under the "Courses" tab.

Free-ed.net (www.free-ed.net/free-ed/healthcare/medterm-v02.asp) offers an online course in medical terminology and CEU credit at reasonable cost.

Des Moines University (www.dmu.edu/medterms) offers a public access, online course in medical terminology.

CRA Courses and Certificate Programs (Including CME Credits)

Association of Clinical Research Professionals (www.acrpnet.org) offers online courses for CME, CNE, and ACRP credit and an industry-recognized CRA certificate program at reasonable cost.

Society of Clinical Research Associates (www.socra.org) offers industry-recognized CRA certification via on-site examination at reasonable cost.

ClinfoSource (www.clinfosource.com) offers online CRA courses for CME and CNE credit and a certificate program at reasonable cost.

Medical Research Management (www.cra-training.com) offers a 140 hour online and classroom CRA certificate program for ACPE credit.

Kriger Research Center International (www.krigerinternational.com) offers online CRA certificate program in multiple languages.

Regulatory Standards and Guidelines

International Conference on Harmonization (www.ich.org) sets international standards for clinical studies of investigational drugs.

International Organization of Standardization (www.iso.org) sets international quality standards recognized by governments for commercial products.

Food and Drug Administration (www.fda.gov) sets safety regulations for foods, drugs, and other medical products in the United States.

European Medicines Agency (www.ema.europa.eu) evaluates and supervises safety regulations of drugs in the European Union.

European Commission/medical devices (http://ec.europa.eu/enterprise/sectors/medical-devices/index_en.htm) sets regulatory standards for medical devices in the European Union.

CRA Professional Organizations

Association of Clinical Research Professionals (www.acrpnet.org) is a global resource for clinical research professionals in the pharmaceutical, biotechnology, and medical device industries and those in hospital, academic medical centers and physician office settings. ACRP provides educational, certification, and networking services for those who support the work of clinical investigations.

Society of Clinical Research Associates (www.socra.org) provides training, continuing education, and an internationally recognized certification program for clinical research professionals.

Profiles of Sponsor and CRO Companies (See detailed list in Chapter 13)

Pharmaceutical Research and Manufacturers of America (www.phrma.org) represents the leading pharmaceutical research and biotechnology companies in the United States.

Biotechnology Industry Organization (www.bio.org), the world's largest biotechnology organization, represents more than 1200 biotechnology companies.

Medical Design & Manufacturing (www.devicelink.com) is an online resource for the medical device industry.

Association of Clinical Research Organizations (www.acrohealth.org) represents the world's leading clinical research organizations.

Salary Surveys for Clinical Research Associates

Applied Clinical Trials Salary Survey (www.appliedclinicaltrialsonline.com) is published annually. Access the most recent survey by entering "salary survey" in the website's Search field.

Association of Clinical Research Professionals CRA Salary Survey (www.acrpnet.org) is published periodically. Access the most recent survey by entering "salary survey" in the website's Search field.

Contract Pharma Salary Survey (www.contractpharma.com) of clinical positions at CROs is published annually. Access the most recent survey by entering "salary survey" in the website's Search field.

Medical Device & Diagnostic Industry Salary Survey (www.devicelink.com) is published periodically. Access the most recent survey by entering "salary survey" in the website's Search field and selecting the Research and Development survey.

Salary.com (www.salary.com) publishes salary ranges for positions in a wide range of industries. Enter "clinical research associate" and if appropriate your targeted zip code in the Salary Wizard.

5

Entering Data Management

A CLINICAL DATA MANAGER IS THE PERSON responsible for processing the clinical study data. This includes tracking, entering, reviewing, cleaning, and coding all the clinical data generated during the study. Ultimately, approval of the medical product for marketing is based on a body of data that shows the product's safety and therapeutic effectiveness. Data integrity therefore is critical, not only to comply with regulatory requirements but also to ensure that accurate conclusions are drawn from the analyzed data. It is the data manager's job to minimize, and if possible eliminate, errors that might cause misinterpretation of and ambiguities in those data.

Some data managers supervise supporting data management personnel, others lead a data management subgroup without supervisory responsibilities, and some oversee outsourced data management services. In all of these cases, the data manager is a member of the clinical study team, serves as a liaison between the team and the subgroup of clinical data management specialists, and is ultimately responsible for the quality of the data.

Maria is the data manager on Nancy's clinical study team. Although she officially reports to a supervisor in the data management department, Maria is accountable to Nancy for all data management activities on the clinical study. Maria and her data management subgroup work for CanDo Pharma, but the clinical team could alternatively contract a contract research organization (CRO) to conduct the study's data management services on its behalf.

The head of Maria's department prepares and maintains an annual budget to cover the data management costs of all scheduled clinical studies. Maria assists the budget planning process by providing estimates of each of her studies' data management costs. Maria's purchasing ability is limited by the expense authority level associated with her job title. However, she can easily receive authorization to negotiate a contract with

BOX 5-1. *Data Management Responsibilities*

- Build and maintain clinical databases
- Develop case report forms
- Create data review guidelines
- Review clinical data for erroneous, missing, or questionable data
- Write and validate error checking diagnostics
- Run data diagnostic programs and special listings
- Create data query conventions
- Produce, track, and resolve queries on problematic clinical data
- Review data tables, graphs, and patient listings
- Coordinate data transfers from third party vendors such as clinical laboratories, medical imaging, and data management CROs
- Review clinical study reports
- Manage data management subgroup

an outside vendor, if CanDo Pharma's data management resources are limited. These vendors might be individual consultants, companies offering computer hardware and software support, or CROs specializing in data management services.

THE ROLE OF THE DATA MANAGER

Maria's interactions on Nancy's study are illustrated in Figure 5-1. She begins her work on the study team by reviewing the draft clinical protocol. She specifically looks at information related to data collection in the sections describing study procedures, treatment schedule, laboratory tests, and clinical assessments. If the protocol's instructions on any of those items are vague, contradictory, or inaccurate, she recommends changes that will clarify data collection.

Maria and her data management subgroup must complete several important tasks during the study start-up phase. Using the clinical protocol as a guide, they write the data management plan (DMP), design and produce the case report forms (CRFs), build the clinical database, and establish the data transfer plan. Once the study is underway, Maria

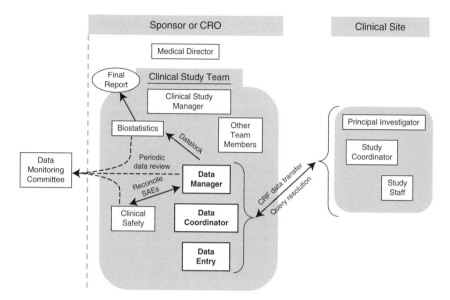

Figure 5-1. Data manager's interactions with the clinical study team.

and her subgroup review incoming data, follow up to correct or clarify questionable entries, and maintain the clinical database. At the end of the study, they ensure that all data have been entered and cleaned before handing the database to the team biostatistician for analysis.

Data Management Plan

The DMP is a document that details the clinical data cleaning and handling procedures. Data collected in Nancy's study include characteristics of the patient (height, weight, age, and previous medical conditions), results of laboratory tests (cholesterol, blood cell count, and X-rays), clinical measurements (blood pressure, bone density, and exercise stress test), the date that each measurement was taken, the type of treatment (drug or placebo), and all observed side effects. For each of these types of data, the DMP specifies the standards for expressing each measurement (e.g., use centimeters not inches), documenting missing data points (e.g., not collected vs. not applicable), and coding medical conditions.

The DMP also gives specific instructions for clarifying and correcting data that appear to be incorrect or ambiguous. The questionable data may be spotted by the clinical research associate (CRA) while monitoring the site or by the data management staff, but only the principal investigators (PIs), or their designees, can make (or authorize) changes to the data. The DMP outlines the procedures for resolving these questions and documenting the PI's changes. Maria's subgroup will follow the instructions in the DMP throughout the study so that all of the collected data, including corrections and updates, are handled consistently.

Case Report Forms Design

In addition to the DMP, Maria also designs the CRFs using the protocol as a guide. In clinical studies, the CRFs serve as a customized laboratory notebook in which the PIs enter the data collected from their patients. Maria designs the CRF pages as a form with check-boxes, multiple-choice lists, yes–no questions, and blank lines/fields for writing comments. She pre-populates some information (e.g., drug name, study number, and visit number) and includes instructions and examples for completing difficult or unusual data fields. She organizes the CRF form to match the data specified in the protocol for collection from each patient. A complete package of CRF pages, called a CRF book, contains all the forms for a single patient for the entire study. In each CRF book, Maria also includes extra forms for recording adverse events that the patient may experience during the study.

For studies that use paper CRFs, Maria compiles the CRF pages in a loose-leaf notebook. When the site coordinators meet with a patient, they take out and complete the appropriate CRF pages as they evaluate the patient. The CRFs are multilayered NCR (no carbon required) pages, allowing the coordinators to make multiple, identical copies as they write on the original form. The clinical site keeps one copy, and the CRAs take a copy to the sponsor for data entry after their monitoring visits.

Nancy's study is using electronic CRFs, so Maria works with her CRF designer to create electronic data capture (EDC) forms. The clinical site staff will enter the data on the EDC forms via a secure web page rather than a paper form. In addition to designing the logical pattern and sequencing of the data fields, Maria specifies rules for entering data into each field. The EDC computer will immediately check the data entries and question data that violate her programmed rules. For example, if the site coordinator tries to enter "10 feet" as the height of a patient, he or she will receive a message asking to confirm the accuracy of the entry and be given an opportunity to change it.

Clinical Database Design

In conjunction with the CRFs, Maria creates the clinical database for Nancy's study. The database is an electronic file that stores all of the data points collected during the study. Most CROs and biopharmaceutical companies use relational databases, which is an electronic method for storing tabular data. For each new clinical study, Maria (or one of her colleagues) programs a unique database structure that matches the protocol's dataset and the fields on the CRF pages.

Maria designs the database so that it is easy for the various team members to use. Site coordinators initially enter the data electronically using EDC. During the study, Maria and her subgroup review the entered data and check for errors and omissions. At the end of the study Tom, the team's biostatistician, analyzes the data. Maria must keep all of these users' requirements in mind as she "builds" the database.

In addition to the database structure and user requirements, Maria programs the database to carry out automated edit checks on the entered data. These are not checks on simple data entry errors but rather on questionable data that require judgment or interpretation. For example, Maria wants to automatically highlight data that are incomplete (missing but not explained), fall outside medically reasonable ranges (but may or may not be accurate), or are inconsistent (i.e., entries in two places that contradict each other). Although some error checking must be done manually, Maria knows that computerized edit checks are more objective, rigorous, and systematic in finding errors in the database.

She also incorporates security features to track everyone who accesses the database and every time they enter, alter, or update a data point. She has issued a unique logon identification code and password to each person authorized to access the database and restricted access to only those data fields that are relevant to their jobs. The security fields that contain information on who, when, and what data changes were made are called "audit trails" and are required by regulatory agencies. Maria will monitor the audit trails during the study to ensure that valid data have been entered by authorized personnel. In addition, regulatory inspectors may review the audit trails during their inspections to confirm that CanDo Pharma is handling clinical data in a secure manner.

Everyone at CanDo Pharma will rely on the database as their sole source of information about Nancy's study. So, Maria thoroughly tests the newly programmed database and confirms that it is functioning properly before she allows any data to be entered. In addition, because Nancy's study is using electronic CRFs, Maria works with David, the CRA, to ensure that the data entry procedures are fully operational and meet the team's requirements. This is called user acceptance testing (UAT). They use realistic but artificial data to test all of the data entry and automated edit check features. All aspects of data handling must be operating correctly before Maria releases the database for use.

On some of her teams, Maria relies on a CRO to build and maintain the clinical database on CanDo Pharma's behalf. In those cases, a data manager at the CRO coordinates the process of designing, programming, and testing the database. Maria uses the CRO data manager as her main point of contact, oversees the CRO's work, but is still ultimately responsible for the quality of the database. Periodically throughout the study and at the end, the CRO transfers the accumulated clinical data to CanDo Pharma's computer system. The CRO and CanDo Pharma must harmonize their actions so that no data are lost or corrupted during the transfers.

Maria specifies all of the rules and requirements for the transfers in a data transfer plan, which is approved by both CanDo Pharma and the CRO, and holds the CRO accountable for all data transfer activities. Because the process of transferring data usually involves reformatting and restructuring the sender's database to conform to Maria's clinical database, the data transfer plan details the steps, responsible individuals, and technical specifications for the transfer to ensure no loss of data integrity. The plan specifies who will conduct the data validation and quality control checks before and during the transfer, as well as details of the data quality standards that must be followed. The plan also states how frequently the data will be transferred, the method of transport (e.g., overnight courier or secure internet connection), the transfer media (e.g., compact disk or internet), the data format, and the procedure for archiving the data transfer account. Maria prepares a similar data transfer plan for other vendors, such as an external laboratory, that will be transmitting data to her clinical database.

Database Entry Training

For Maria, the final study start-up step is training those who will enter the data. Because Nancy's study uses EDC, David and Maria train the study coordinators on data entry procedures as part of the investigator meeting.

(For studies that do not have an investigator meeting, Maria and the CRA would arrange for training at each clinical site.) After the study coordinators complete their training, Maria authorizes their access to the database by issuing their unique logon IDs and passwords.

For studies that use paper CRFs, Maria relies on her subgroup to enter the data. (If the database activities are outsourced, the CRO would perform the data entry.) She provides them with an annotated CRF document, which gives instructions and examples for correctly entering each data point from the paper CRF into the database. For added accuracy, CanDo Pharma's standard operating procedures (SOPs) require "double data entry." That is, two different people enter the same data from each CRF, and the database only accepts the data if the two entries match. If they do not match, the difference must be resolved and explained before the database accepts it.

The nonnumeric information reported in the comment fields of the CRFs must be entered into the clinical database using standardized definitions of medical terms. By coding these terms in a consistent manner, Maria can sort the clinical data into the appropriate categories for later analysis. The Medical Dictionary for Regulatory Activities (MedDRA) is one industry standard for medical data coding.

Data Cleaning

Throughout the study, Maria and her subgroup review and "clean" the data as they receive the EDC entries from the sites. The data must be complete, accurate, consistent, and quantifiable. Maria has programmed the database to automatically reject or highlight questionable data entries, which the site coordinators must correct immediately. Other errors are caught through the automated edit checks. However, computer programmers cannot predict all of the possible errors, and some edit checking relies heavily on human judgment. To identify such errors, Maria's subgroup performs manual edit checks.

Maria must resolve all inconsistencies and suspected errors in the database. These take many forms: A positive pregnancy test in a male patient, a temporary medical condition that lists no end date, a treatment-related adverse event with a start date before treatment occurred, or an observation listed on the physical examination CRF page (for example, a skin rash) that is not also listed on the adverse event CRF page.

When Maria and her subgroup identify discrepancies or questionable data, they must resolve them following the procedures outlined in the DMP. Only the PIs (or their designees) may authorize additions, changes, and updates to the clinical data. Maria's subgroup therefore consults the clinical sites in the form of a clinical data query. Maria writes the query, referencing the specific CRF item, explaining why the entry appears incorrect, and asks the PI to respond. In most cases, the PI can quickly clarify the entry with either an explanation or new information. When she receives the completed, signed, and dated query response from the PI, Maria makes the appropriate changes to the database and indicates that the issue is officially resolved. Using an EDC system, query resolution is usually rapid and ongoing, because the query, response, and database updates are all communicated electronically over a secure internet connection between the clinical site and CanDo Pharma.

In some cases, the clinical circumstances surrounding the flagged entry may be more complex and require discussions to resolve the query. The PI, the CanDo Pharma physicians, and perhaps others discuss these special circumstances and decide how to capture the information accurately in the database. Maria then updates the database accordingly and documents the rationale for the final, entered data.

In other cases, the clinical sites may discover errors in their previous data entries, or David, the CRA, may find errors when he is monitoring the sites. They submit these site-generated corrections to Maria with authorization from the PI to correct or update the database.

In addition to managing the study database and resolving queries, Maria supports the work of the data monitoring committee (DMC). (See Chapter 6 for details on DMC operations.) Before each scheduled DMC meeting, Maria coordinates her work with CanDo Pharma's liaison biostatistician to prepare summaries of the accumulated study data for the DMC meeting. Maria's responsibility is to ensure that the data that will be extracted for DMC review have been properly reviewed and cleaned.

At the end of the study, Maria and her subgroup systematically review the database one final time to check for any overlooked errors. With EDC, Maria has cleaned most of the data as they were entered, leaving few unresolved queries. With paper CRFs, Maria and her subgroup receive the CRFs and enter the data at a slower rate, and query resolution is often more laborious. In either case, after Maria's subgroup has resolved all of the queries and she is certain that all data-related issues have been resolved, she "locks" the database. Database lock means that she will permit no further alterations to the data. After datalock, the data are unblinded, and Nancy's study team finally learns which treatment was given to each patient. Maria submits the locked and unblinded database to Tom, the biostatistician, for data analysis.

Although others on the team take primary responsibility for writing the clinical study report, Maria supports their efforts. In particular, she works with Tom, the biostatistician, as he generates tables, listings, and graphs of the data. (See Chapter 6 for details of clinical data analysis.)

During the report approval process, Maria reviews the final report, carefully checking all of the data listings, tables, and graphs. As the data manager, Maria must confirm that all of the data represented in the report are correct and consistent with the database. Her signature is among the most important of all the team members, because everyone relies on her to verify the data's integrity.

ONE OF MARIA'S DAYS

While Maria finishes her morning coffee, she reviews her tasks for today: attending a study team meeting to schedule an interim analysis, reviewing a draft study protocol, and writing the data management plan for a new study.

Maria serves as the data manager for all of the clinical studies in Dr. Abernathy's development program for the new CanDo Pharma drug, and each study is at a different stage. This morning, one of those study teams is meeting to finalize the plans for an interim analysis. (See Chapter 6 for more details about interim analyses.) The study has now met the pro-tocol requirement, which specified an interim analysis after half of the patients were enrolled. Because the patients are being treated for two years, less data have been collected on the most recently enrolled patients than those who enrolled earlier. The team wants to include as much data as possible in the interim analysis, so the CRAs have been working hard to encourage all of the clinical sites to enter their data from the enrolled patients as quickly as possible.

For Maria, the most important issue is agreeing on the cut-off date for data that will be included in the interim analysis. Maria will concen-trate on reviewing and querying CRF data received before that date. The team agrees to set the end of the month as the cut-off date.

One important factor in the team's decision for setting the cut-off date is the transfer of data from an external laboratory. According to

CanDo Pharma's agreement with the laboratory, data for this study are electronically transferred to Maria's clinical study database quarterly, and the next scheduled transfer is this month. Results from the laboratory will be an important aspect of the interim analysis, and the team agreed to delay the interim analysis until the latest laboratory data have been received.

After the team meeting, Maria reviews the draft protocol for a new clinical study that Dr. Abernathy is planning with the CanDo Pharma drug. She has attended meetings with the newly created clinical study team where this protocol has been discussed, so she is already familiar with the study's design and objectives. From a data management perspective, Maria only spots one difficulty in the draft. Several liver function tests are listed as secondary end points in the introductory section, but the methods section, which details the procedures and patient assessments, does not mention any liver function tests. This discrepancy is probably an oversight, but it is an important issue for Maria. She needs to know whether to include fields in the CRF pages and in the clinical study database for collecting and storing the liver test results. She sends her comments on the protocol, including this apparent discrepancy, to the study manager by email just before lunch.

Maria plans to spend the afternoon writing the data management plan for another clinical study of the CanDo Pharma drug. Dr. Abernathy approved the final protocol for this study last week and asked the clinical study team to start the study as soon as possible. It is similar to other studies in the drug's development program, including Nancy's study, and Maria can draw on her earlier data management plans to prepare the new DMP.

However, before she can make much progress, Maria receives a call from the manager at the contract laboratory. He informs her that the laboratory's analysis schedule has been disrupted by a series of power outages. He has already reassured the CanDo Pharma study manager that no samples were harmed, but the laboratory analyses have been delayed, and he is concerned about meeting the deadline for the quarterly data transfer to Maria's database. The laboratory's standard procedures require him to conduct a series of performance tests to confirm that the laboratory's computer is working properly before he authorizes any new data transfers. He estimates that the data transfer will be delayed until sometime next month.

Maria explains the importance of transferring the data as originally scheduled, and she offers a number of possible solutions. However, the laboratory manager says he has already tried those solutions, and none were successful. Furthermore, he cannot predict whether the performance tests will uncover other difficulties, which may also need to be addressed.

They both know that quality cannot be compromised. The laboratory data will be useless if they cannot control the data transfer and ensure that no errors were made during transmission. Given the importance of the interim analysis timetable, Maria recommends that they hold an emergency teleconference tomorrow morning to discuss the problem and agree on a plan. Maria will include the key people from the CanDo Pharma study team, the manager will include the analysts from the laboratory, and they will both invite their respective information technology experts to participate.

Maria asks one of her data management associates to handle the arrangements for scheduling the teleconference. She then places calls to the study manager and Dr. Abernathy to explain the problem in detail and personally invites them to participate in the teleconference. She is still confident that together they can find a way to keep the end-of-month cut-off date, but the new plan will require input from everyone who is involved with the interim analysis.

Before tomorrow's teleconference, Maria promises to gather as much information as possible for the team's consideration. Based on the enrollment dates for each patient, the protocol's schedule of required laboratory tests, and the date of the last data transfer from the external laboratory, Maria can compile a summary of the new data points she expects to receive from the laboratory in the upcoming data transfer.

While she is preparing the summary, Maria receives a call from David, the CRA on Nancy's clinical study. He is at Dr. Chase's clinical site to conduct routine monitoring and has been trying to help Sally with a data entry question. One of the study patients has been undergoing acupuncture treatment for a condition that is unrelated to the study. David knows that acupuncture treatment does not disqualify the patient, but this information must be recorded in the study database. The electronic CRFs include a page to enter concomitant medications (i.e., off-study drugs that the patient is taking), but Sally and David are unsure where to enter the acupuncture information. Over the phone, Maria gives instructions, step by step, while Sally scans through the electronic CRFs and enters the acupuncture details on the appropriate page.

Maria then continues preparing the summary of information about the laboratory data transfer. Her associate stops by to confirm that a CanDo Pharma meeting room has been reserved, the teleconference communications have been arranged, and all participants have accepted their meeting invitations. Maria wants to send her summary by e-mail to all of the external laboratory and CanDo Pharma participants today, so that they can review it before the meeting if they wish. Although it is already late afternoon, she will stay at work until she finishes it. Hopefully, the meeting will go well tomorrow and she can then return to writing the data management plan, which is still waiting on the corner of her desk.

HOW MARIA GOT HER DATA MANAGER JOB

Requirements

After Maria received her bachelor's degree in biology, she took a job as office assistant for a physician at a major medical center. Her duties included retrieving and filing patient charts, scheduling medical procedures, and

---

BOX 5-2. *Requirements for a Clinical Data Management Position*

- At a minimum, bachelor's degree in science or a license in nursing, pharmacy, medical technology, or physician's assistant
- Clinical experience: 1 to 3 years of clinical data experience in a medically related field (e.g., study coordinator)
- Knowledge of medical terminology, pharmacology, anatomy, and physiology
- Knowledge of CDISC and MedDRA conventions
- Knowledge of GCPs and ICH regulatory requirements
- Proficient with database management systems and relational databases
- Familiarity with electronic applications such as EDC and IVRS
- SAS programming is helpful but not required
- Detail oriented and excellent organizational skills
- Good interpersonal and communication skills, written and verbal
- Ability to interact confidently and cooperatively with other team members
- Ability to meet deadlines

submitting insurance information. Through this experience, Maria learned medical terms and how to use several medical computer systems.

She enjoyed working with computers and manipulating sets of data. Because she had conducted an independent research project as an undergraduate, Maria was familiar with organizing scientific data, using spreadsheets, and conducting simple statistical analyses. A group of physicians at the medical center therefore asked her to organize the data for several independent clinical studies that they wanted to analyze and subsequently publish. This assignment pushed Maria to learn more about the structure and programming of relational databases. Her success in completing this assignment led to other data handling opportunities.

Among those additional opportunities was her first exposure to industry-sponsored clinical studies. The medical center served as the clinical site for several experimental drug studies, and one study coordinator asked Maria to help with the recordkeeping. Under the coordinator's direction, Maria recorded patient histories, filed laboratory results, and completed CRFs.

Several of the clinical studies used EDC and Interactive Voice Response System (IVRS). (See Chapter 6 for details on IVRS.) Maria welcomed the opportunity to learn about electronic data entry and the clinical databases used by sponsor companies and CROs, including the procedures for electronic queries and datalock. She liked the challenge of handling clinical data and using computer tools to manage the datasets. Through discussions with the CRAs who monitored her site, Maria realized that a career in clinical data management was a good move for her. In data management, she would be able to work in information technology and still directly handle clinical data.

Finding a Clinical Data Management Position

You can find entry-level data management positions at sponsor companies and at CROs. (See Chapter 13 for a list of industry websites.) Most of these companies post job openings on their websites, which can easily be searched by job title. The job titles for entry-level data management positions vary considerably between companies, but look for titles such as data coordinator, clinical data associate, clinical data specialist, or data analyst.

For entry-level data management candidates, CROs offer several advantages over sponsor companies. First, CROs offer more structured training than sponsor companies and set measurable skill standards. Second, entry-level workers at a CRO quickly gain experience on many different clinical databases, because each of the CRO's clients sets its own specifications for database structure and data handling procedures. Third, because of the variety of studies and therapeutic areas represented by the CRO's clients, entry-level workers quickly gain wide experience on all stages of database management (i.e., start-up, maintenance, and datalock) and with many disease conditions. Finally, because many sponsors now outsource routine data management tasks (such as data entry and query production), CROs are the best place to learn all of the procedures involved with managing clinical data.

Large companies offer better opportunities for an entry-level worker than small companies. Large sponsor and CRO companies typically have full-function data management departments. The experienced staff, established procedures, and validated computer systems offer an entry-level candidate the best opportunities for gaining high-quality training, experience, and career advancement. By contrast, small CROs have limited resources and typically hire only experienced data managers, who are expected to perform multiple duties. Similarly, small sponsor companies also have limited resources and typically outsource most of their data management activities to a CRO; the data manager is highly experienced but merely oversees the work of the contracted CRO and rarely deals with the data directly.

Landing the Job

The hiring managers at both CROs and sponsor companies look for the same skills and qualities in their job candidates: a basic understanding of clinical studies and experience with computer spreadsheets, relational databases, and data analysis.

Maria's independent study project in college and her work in the physician's office gave her the necessary qualifications for an entry-level data management position, but there are other ways to qualify for this position. Those with a nursing or pharmacy background and an aptitude for computers can also qualify for data management positions. Study coordinators, who are familiar with reading clinical protocols, following good clinical practices (GCPs), completing CRFs, and resolving data queries, also have excellent prerequisites for data management jobs. (See Chapter 3

BOX 5-3. *Tips for Getting a Data Manager Position*

- Gain expertise with spreadsheets and relational databases
- Gain experience with organizing scientific data (e.g., independent study projects, laboratory assistant, or gathering survey data)
- Consider volunteering at a medical center or doctor's office
- Study coordinator experience is a big plus
- Take online courses on GCP and ICH
- Learn medical terminology
- Learn MedDRA and CDISC coding
- Consider obtaining SAS programming certification
- Consider SQL training
- Join a professional society such as SCDM or DIA
- Search data management job postings on websites of drug, medical device, and CRO companies
- Preferentially look for entry-level jobs at large CROs and biopharmaceutical companies

for details on landing a study coordinator job.) Alternatively, those with a degree in computer science and an interest in life science may also qualify.

All of the hiring managers ranked Maria as a highly desirable data management candidate. Although she had held no data management positions, she was familiar with medical procedures and clinical data. She understood GCPs, and she knew how to enter data on paper CRFs and in EDC. Furthermore, the CRAs who monitored her clinical site confirmed that Maria submitted high-quality CRFs and cooperated with them to resolve data queries quickly.

The hiring managers were less impressed with the candidates who competed with Maria for entry-level positions, even though some were study coordinators. Data management workers at all levels must adhere to strict requirements when handling data and tenaciously work toward finalizing a high-quality database. The hiring managers therefore rejected candidates who submitted résumés with grammatical and spelling errors and candidates who, as study coordinators, had a history of submitting incomplete and incorrect CRFs.

Because data managers work closely with the rest of the study team throughout the study to resolve questions about CRF design, database

structure, and the entered data, data managers must have excellent interpersonal skills. Hiring managers, therefore, eliminated candidates who
did not speak clearly and professionally during their interviews or who
had a history of being disrespectful in stressful situations.

Working for a Sponsor Company Versus a CRO

Although Maria could have taken an entry-level data management position at a sponsor company, she decided to work for a large CRO. Her first
position at the CRO was as a clinical data coordinator. Under the supervision of a data manager, Maria successfully completed the CRO's
required data management training modules, including International
Conference on Harmonization (ICH) standards and the CRO's SOPs for
conducting data management activities. She also learned how to enter
clinical study data using the CRO's computer system.

Initially, Maria assisted her coworkers and worked with clinical databases
that were already built for clinical studies that were already in progress.
After completing study-specific training to learn about the clinical protocol,
experimental drug, and the disease being treated, Maria was allowed to
work more independently in reviewing the CRF data entries for possible
errors. She conducted manual and automated edit checks, wrote queries,
and tracked the sites' responses until the questionable data were satisfactorily resolved. At the end of the studies, she learned how to proofread the
tables, graphs, and patient listings attached to clinical study reports.

As Maria gained on-the-job experience, she was given more responsibility, particularly during the start up of new clinical studies. She designed
CRFs, created data review guidelines, defined the automated edit checks,
and decided which standard dictionary (e.g., MedDRA) would be used
for coding medical terms.

During this time, Maria joined a professional society for data managers. (The Society for Clinical Data Management [SCDM], the
Association of Clinical Research Professionals [ACRP], and the Drug
Information Association [DIA] are three prominent societies for data
managers.) The advantages of membership include training classes, an
official clinical data management (CDM) certification program, newsletters and other data management-oriented publications, an annual society
meeting, industry networking, and knowledge sharing on the latest technologies and data management trends.

The two most effective ways for Maria to advance her career in data management are to gain technical experience with clinical data management systems and maintain her listing in her professional society directory. The database systems used to manage clinical data are constantly being enhanced, refined, and upgraded to meet the requirements of regulatory agencies and the database users. The senior management of sponsor companies and CROs rely on their data management experts to evaluate and make recommendations on software and hardware improvements, which often require large, capital investments. In order for Maria to reach an influential data management position, she needs to demonstrate, through experience and training, her expertise with industry standards such as Clinical Data Interchange Standards Consortium (CDISC), Structured Query Language (SQL), and SAS.

CDISC is a global nonprofit organization that sets platform-independent data standards for acquisition, exchange, submission, and archiving clinical data. Regulatory agencies are increasingly encouraging sponsors to use CDISC standards for classifying, reporting, and archiving clinical data. SQL is a database computer language designed for managing data in relational database management systems. SQL software features include data querying, data updating, and data access control. SAS is an integrated system of software products provided by SAS Institute and offers applications such as data management, statistical analysis, and data warehousing. SAS is a validated software system recognized by many regulatory agencies.

Listing expertise with industry-standard data management systems in her résumé gives Maria an important advantage over other candidates if she decides to apply for senior data management positions at other companies. Maria's listing in the society's membership directory also makes her visible to recruiters and hiring managers who are looking for experienced data managers.

If she chooses to stay within her company, whether it is a sponsor or CRO, her career advancement will depend mostly on her on-the-job performance, her technical expertise with data management systems, good leadership skills, and her ability to handle critical tasks that involve tight deadlines and creative problem solving.

Maria performed well and could have advanced her career to more responsible data management positions at the CRO. Data managers at CROs have broad experience with clinical databases in multiple therapeutic areas and supervise a staff of workers. As a CRO data manager, Maria would be the primary point of contact for the sponsors who outsource their data management work. In addition to overseeing the clinical database activities, she would also train new data management staff, manage work assignments, and establish new and revised data handling procedures.

Although Maria saw the advantages of continuing to work at the CRO, she wanted experience at a company that developed its own products. In addition, the compensation and benefits offered by sponsor companies are often more attractive than at CROs. Because Maria was listed in the membership directory of her data management society, the CanDo Pharma hiring manager easily identified her as a potential candidate. All of Maria's experience as a senior-level clinical data analyst at the CRO made her an attractive candidate as a data manager at CanDo Pharma.

After receiving leadership and supervisory training at CanDo Pharma, Maria became a data manager with responsibilities for managing both clinical databases and a small support staff. By the time Nancy's study started, Maria had served as the data manager on several clinical studies in Dr. Abernathy's drug development program. She was fully prepared and easily incorporated Nancy's study into that cluster of clinical studies on the CanDo Pharma drug.

MOVING FORWARD

Because Maria liked handing clinical data, she decided to enhance her data management skills, progressing from an entry-level data coordinator at the CRO to her current data manager position at CanDo Pharma. Although the titles of data management positions (e.g., clinical data specialist, clinical data analyst, and clinical data associate) vary considerably, most sponsor companies and CROs offer a career track in data management. From an entry-level position, data management workers typically advance their level of responsibility every 3–5 years, depending on performance. Workers like Maria reach data manager positions after gaining experience at two or three lower levels.

Maria could continue her data management career through promotions to senior data manager and the director level. Eventually, she might qualify to head a data management department and be responsible for all of the clinical data management staff and its activities, either at CanDo

Pharma, another sponsor company, or a CRO. On the other hand, Maria might choose to specialize in one area of data management, such as CRF design, database programming, or data management systems.

CRF designers work with data managers and the clinical study teams to prepare customized CRF pages for each clinical study. They must have an aptitude for creative design and desktop publishing software, as well as a thorough understanding of the types of data that are collected in clinical studies. If the final CRFs are paper copies, the CRF designer arranges for printing the CRF books and oversees the work of the printing vendor. If the final CRFs reside in an EDC system, the CRF designer programs the electronic CRF pages and works with the data manager and the study team, who conduct the UAT. Together, they ensure that the data entry and edit check procedures work properly.

Database programmers set up and maintain the customized clinical databases for each clinical study. Most biopharmaceutical programmers are proficient with SAS software, which is a validated programming system for data mining and tabulation, and SQL software, which is used for querying and managing data. In building the database, programmers must comply with numerous regulatory requirements regarding data security. They conduct tests of the database to ensure accurate data entry and storage, restrict access to the data through secure password procedures, and establish audit trails for tracking each person's interaction with the database. Programmers also write computer programs to extract data from the database and organize it in the customized tables, graphs, and listings. Certification in SAS programming or SQL gives data management job applicants a significant advantage over other applicants.

Large data management departments include a separate support group that is responsible for maintaining the database computer software and hardware systems. Regulatory agencies require accurate collection and safe storage of all clinical data. Data systems managers, therefore, must have strong computer skills and a thorough understanding of relational databases. They must also comply with regulatory requirements for computer validation and data integrity, the use of electronic signatures, and GCPs.

Rather than data management, Maria might choose to advance her career in another direction. If she wishes to apply her data management experience toward broader study management, she could easily move into the project management function, where she would be responsible for managing an interdisciplinary team that oversees a program of clinical studies. (See Chapter 11 for more details on project manager responsibilities.)

Alternatively, Maria's data management skills, understanding of clinical protocols, and experience in solving the problems associated with data collection make her highly qualified to monitor clinical sites. She may wish to pursue a CRA position. (See Chapter 4 for details of CRA positions.) After working as a CRA, Maria could rapidly qualify for a position as a clinical study manager. (See Chapter 11 for more information about study manager duties.)

Maria's experience in resolving discrepancies in clinical databases is also excellent training for clinical quality assurance (CQA) positions. Key skills needed by CQA auditors are attention to detail and a thorough understanding of GCPs. If Maria is interested in moving toward a career in quality systems and regulatory activities, she could easily enter this career track as a CQA auditor. (See Chapter 7 for more information about CQA auditor activities.)

Whether Maria chooses to continue working in clinical data management or move to the project manager, clinical study manager, or CQA career tracks, she can advance to positions that offer higher salaries, greater company visibility, more decision-making authority, budget responsibilities, and greater corporate benefits such as bonus and stock option packages. Although these senior level positions will move her further away from day-to-day clinical activities, Maria's work would still revolve around clinical studies by contributing to the strategic planning, oversight, and decisions regarding new medical product development.

DATA MANAGEMENT RESOURCES

Good Clinical Practices Training and Certificate Programs

ClinfoSource (www.clinfosource.com) offers online GCP courses and certification for CME and CNE credit at reasonable cost.

Kriger Research Center International (www.krigerinternational.com) offers online GCP courses and certification in multiple languages on the country-specific web pages under the "Courses" tab.

Association of Clinical Research Professionals (www.acrpnet.org) offers online GCP courses and certification for CME, CNE, and ACRP credit at reasonable cost.

GCP Training Online (www.gcptraining.org.uk) offers online GCP training and certification, emphasizing the European Clinical Trial Directive, at reasonable cost.

Institute of Clinical Research (www.instituteofclinicalresearch.com) offers online GCP and European Clinical Trial Directive courses and certification at reasonable cost.

Data Management Courses and Certificate Programs (Including CME)

Society for Clinical Data Management (www.scdm.org) offers online courses for IACET and CEU credit and a certified clinical data manager (CCDM) certification program.

Kriger Research Center International (www.krigerinternational.com) offers online clinical data management courses and a certificate program in good clinical data management practices in multiple languages on the country-specific web pages under the "Courses" tab.

ClinfoSource (www.clinfosource.com) offers online clinical data management courses for CME and CNE credit and a certificate program at reasonable cost.

SAS programming (www.sas.com/certify) offers online courses and a global certification program in SAS programming.

Regulatory Standards and Guidelines

International Conference on Harmonization (www.ich.org) sets international standards for clinical studies of investigational drugs.

International Organization of Standardization (www.iso.org) sets international quality standards recognized by governments for commercial products.

European Medicines Agency (www.ema.europa.eu) evaluates and supervises safety regulations of drugs in the European Union.

European Commission/medical devices (http://ec.europa.eu/enterprise/sectors/medical-devices/index_en.htm) sets regulatory standards for medical devices in the European Union.

Food and Drug Administration (www.fda.gov) sets safety regulations for foods, drugs, and other medical products in the United States.

Code of Federal Regulations, especially CFR 21, part 11, subparts A–C (www.gpoaccess.gov/CFR) stipulates the regulatory standards for electronic records and electronic signatures.

Medical Coding, Dictionaries, and Terminology

Medical Dictionary for Regulatory Activities (www.meddramsso.com) is a clinically validated dictionary of terms used to report adverse event data from clinical studies.

Clinical Data Interchange Standards Consortium (www.cdisc.org) sets clinical data conventions for acquiring, exchanging, submitting, and archiving clinical data.

ClinfoSource (www.clinfosource.com) offers an online medical terminology course for CME and CNE credit at reasonable cost.

Kriger Research Center International (www.krigerinternational.com) offers an online course in medical terminology in multiple languages on the country-specific web pages under the "Courses" tab.

Free-ed.net (www.free-ed.net/free-ed/healthcare/medterm-v02.asp) offers an online course in medical terminology and CEU credit at reasonable cost.

Des Moines University (www.dmu.edu/medterms) offers a public access, online course in medical terminology.

Data Management Professional Organizations

Society for Clinical Data Management (www.scdm.org) is an organization that fosters educational, certification, and networking services for clinical data management professionals.

Association of Clinical Research Professionals (www.acrpnet.org) is a global resource for clinical research professionals in the pharmaceutical, biotechnology, and medical device industries. ACRP provides educational, certification, and networking services for those who support the work of clinical investigations.

Drug Information Association (www.diahome.org) is an international professional association for those who are involved in discovery, development, regulation, surveillance, or marketing of biopharmaceutical products.

Profiles of Sponsor and CRO Companies (*See detailed list in Chapter 13*)

Pharmaceutical Research and Manufacturers of America (www.phrma.org) represents the leading pharmaceutical research and biotechnology companies in the United States.

Biotechnology Industry Organization (www.bio.org), the world's largest biotechnology organization, represents more than 1200 biotechnology companies.

Medical Design & Manufacturing (www.devicelink.com) is an online resource for the medical device industry.

Association of Clinical Research Organizations (www.acrohealth.org) represents the world's leading clinical research organizations.

Salary Surveys for Clinical Data Managers

Applied Clinical Trials Salary Survey (www.appliedclinicaltrialsonline.com) is published annually. Access the most recent survey by entering "salary survey" in the website's Search field.

Medical Device & Diagnostic Industry Salary Survey (www.devicelink.com) is published periodically. Access the most recent survey by entering "salary survey" in the website's Search field and selecting the Research and Development survey.

Salary.com (www.salary.com) publishes salary ranges for positions in a wide range of industries. Enter "clinical data specialist" and if appropriate your targeted zip code in the Salary Wizard.

6

Entering as a Biostatistician

A BIOSTATISTICIAN IS THE PERSON RESPONSIBLE for all statistics-related activities associated with designing, analyzing, and interpreting the data from clinical studies. Biostatisticians play a critical role in product development, because regulatory agencies will only approve a medical product if the clinical data meet rigorous statistical criteria for demonstrating safety and efficacy. Sponsors, therefore, are diligent in designing clinical studies so that they collect the right data for analysis of safety and effectiveness, and they rely on biostatisticians to guide them.

Tom is the biostatistician on Nancy's clinical study team. Although he reports to a supervisor in CanDo Pharma's biostatistics department, Tom is accountable to Nancy for all biostatistics activities on their clinical study. The study teams at CanDo Pharma prefer to use their own biostatisticians, but they may contract the services of biostatisticians at a contract research organization (CRO), depending on internal workloads and priorities.

THE ROLE OF THE BIOSTATISTICIAN

The interactions of Tom and his colleagues on Nancy's study team are illustrated in Figure 6-1. Study biostatisticians typically do not have supervisory responsibilities, but Tom can rely on clerical and administrative support from his department. CanDo Pharma also employs statistical programmers to assist the biostatisticians in writing the code for their statistical analyses.

The head of Tom's department prepares and maintains an annual budget to cover the statistical analysis costs of all scheduled clinical studies and other departmental expenses. Typically, these expenses are modest

---

BOX 6-1. *Biostatistician Responsibilities*

- Create statistical text for protocols (study design, end points, sample size, methods for analysis)
- Follow the statistical aspects of GCP and relevant SOPs
- Prepare statistical analysis plans
- Review and approve key study-related documents (CRF design, data management plan)
- Set up randomization specifications for patient assignments
- Specify shells for patient listings, tables, and graphs for clinical study reports
- Write, test, validate, and execute software programs to produce datasets and tables, listings, and graphs for inclusion in clinical reports
- Perform, present, and interpret statistical analyses
- Derive statistical conclusions and make recommendations accordingly
- Support and facilitate DMC activities
- Facilitate DMC charter and member selection
- Coauthor and review statistical section of clinical reports
- Archive statistical programs
- Integrate data sets for market authorization applications (MAAs)
- Prepare integrated summaries for MAAs (summary of clinical efficacy, summary of clinical safety)

---

and mainly consist of the cost of purchasing and maintaining statistical software and hardware systems. Tom's purchasing ability is limited by the expense authority level associated with his job title. However, he can easily receive authorization to negotiate a contract with a CRO, if his department decides to outsource the statistical services for a study. In those cases, Tom would oversee the work of the CRO biostatisticians and ensure that they meet CanDo Pharma's requirements for analytical methods and quality.

Study Design

Long before the protocol for Nancy's study is written, Tom actively contributes to Dr. Abernathy's strategy team discussions regarding study

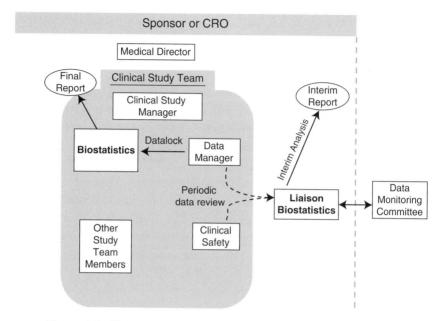

Figure 6-1. Biostatatistician interactions with the clinical study team.

design. In order for Tom to analyze the data properly and draw meaningful conclusions, the study team must collect the right amount and the right kinds of data. Throughout their discussions, the strategy team balances clinical and statistical considerations. Dr. Abernathy wants to include patients, procedures, and data to meet his clinical objectives, but those considerations may not be statistically reasonable. Tom wants the protocol to include data collection requirements that ease his ability to conduct the statistical analysis, but such data collection may not be clinically feasible.

When the team agrees on the study's objectives and final design, Tom then specifies the statistical requirements to meet those objectives including the randomization procedures and the choice of statistical tests. Tom realizes that the other team members are not statistics experts, and he is careful to explain his rationale and statistical choices in nontechnical language.

Tom writes an important section of the protocol, describing the statistical methods he will use to analyze the data. International Conference on Harmonization (ICH) standards require Tom to include certain statistical details, such as the sample size rationale, end point definitions, operating characteristics (e.g., false positive rate, power, randomization ratio, and sample size), endpoint analysis methods, study termination criteria, data accounting procedures, analysis plan deviation procedures, and criteria for selecting patients for analysis. He also reviews other protocol sections (such as the objectives, synopsis, and data collection requirements) to make sure they are consistent with his text.

Tom and his teammates make provisions in the protocol for stopping the study early under certain clinical situations. Their primary concern is patient safety, and they will stop the study if the results indicate that the treatment is doing some unanticipated harm. However, they may also wish to stop the study early if the treatment is overwhelmingly beneficial or the study is futile (i.e., the early data indicate that it is unlikely they will be able to draw definitive conclusions from the study). Tom incorporates these clinical considerations into defined "stopping rules." In the statistics section of the protocol, he justifies the criteria for stopping the study and describes the statistical methods he will use to analyze the incomplete dataset of a prematurely terminated study.

Tom knows that despite everyone's best efforts, data collection will not be perfect. When the study is completed, some data points will probably be missing. Other data points will be collected but cannot be used because of a known problem during treatment of the patient. (For example, a blood sample was drawn at an incorrect time.) Tom may also choose not to analyze some of the data, even though they were collected correctly. In the protocol, Tom justifies the procedures he intends to use for selecting and rejecting individual data points and explains how he will handle missing and questionable data.

In addition to decisions about individual data points, Tom decides, in conjunction with the team, which patients he will include in the analysis. They expect that some patients will not complete the study as outlined in the protocol. In some cases, patients meet all of the inclusion and exclusion criteria but decide not to participate in the study (i.e., eligible but not enrolled). Some patients may be assigned to a treatment group but withdraw before they are dosed (i.e., randomized but not treated). Some patients are dosed but drop out of the study before completing treatment (i.e., treated but not completed). In the statistics section of the protocol, Tom states which groups of patients he will include in his analysis: All eligible, all randomized, all dosed, or only those who completed the study as specified (i.e., all evaluable). At the end of the study, but before he begins his analysis, Tom will review the blinded data listings and assign each patient to one of these predefined analysis groups.

Study Start Up

After the protocol is finalized, Tom contributes to start-up activities such as database structure definition, case report form (CRF) review, data edit specifications, and randomization specifications. He also writes a statistical analysis plan (SAP), which is required by the regulatory authorities and CanDo Pharma's standard operating procedures (SOPs).

Tom works with Maria, the data manager, to plan all of the data collection activities. To facilitate his data analysis, they design the CRF pages and the database structure, define the structure of the database tables, and establish the links of key data fields between the tables. They also agree on the format for each data field (e.g., numerical vs. text) and the edit check procedures. This advanced planning makes it easier for Tom to extract and analyze the data at the end of the study and contributes to the quality of the final dataset. (See Chapter 5 for more details on CRFs and clinical databases.)

With input from Nancy, the study manager, Tom develops the randomization specifications. Each patient must be assigned randomly to either the drug or control treatment group. Tom uses statistical methods to generate the randomization scheme that stratifies the assignments, taking into account the various clinical sites in the three countries where the study is conducted. Because Nancy's study is blinded, the treatment assignments are coded (usually A, B, C, etc.) to conceal the identity of the placebo and drug treatments. Tom's specifications are then incorporated into an automated treatment assignment system called interactive voice recognition system (IVRS). When a site enrolls a new patient in the study, the study coordinator contacts IVRS by telephone. IVRS applies the randomization scheme to assign the patient to one of the blinded treatment groups and instructs the study coordinator to issue the corresponding coded treatment supplies.

Statistical Analysis Plan

In the statistical analysis plan (SAP), Tom elaborates on the statistical methods that he described in the protocol. As in the protocol, he states the study's objective in testable, statistical terms. His analysis depends on clear SAP definitions for selecting which patients, clinical measurements, and statistical methods he will use to test whether the study's objectives were achieved. In the SAP, he writes detailed specifications for all of those statistical choices. Tom also includes in his SAP detailed statistical procedures for analysis of data subsets required by ICH and other regulatory bodies; these include subgroup analyses based on age, race, gender, pretreatment disease status, coexisting medical conditions, and concomitant medications (i.e., drugs, other than the experimental treatments, taken by the patient during the study).

Tom must write the SAP before he begins analyzing the data. Otherwise, he might be tempted to choose analysis procedures that make the results look better than they really are. If the team wishes to assess the results before the study is completed, Tom incorporates "interim analysis" procedures in the protocol and SAP for analyzing the incomplete database. Because Nancy's study is blinded, Tom must state clear rules in the SAP for handling the interim analysis data so that disclosure of these results does not influence how the remaining part of the study is conducted and analyzed.

Tom cannot foresee all of the possible difficulties that might occur when he begins analyzing the final data. Anticipating that he might need to revise or deviate from the procedures in his SAP, Tom includes procedures in the study protocol for documenting all deviations from his planned analysis.

Data Monitoring Committee

During the conduct of clinical studies, biostatisticians support the data monitoring committees (DMCs). The DMC is an independent committee established by the sponsor to oversee a study that involves treating seriously ill patients or potentially hazardous products. (See Chapter 2 for details of DMC regulatory requirements.) The DMC members include physicians, biostatisticians, and other medical experts who are paid for their services but are not employees of the sponsoring company. The DMC meets at regular intervals to assess the progress of the study, review the safety (and possibly efficacy) data, and recommend whether to continue, modify, or stop the study. It operates under a DMC charter, which contains the rules for open and closed DMC meetings, requirements for issuing DMC reports, procedures for communicating and implementing DMC recommendations, and confidentiality requirements.

Because the DMC operates independently from both the sponsor and the study team, it can view unblinded data without biasing the study and render decisions that are not influenced by the sponsor's executives or teams. Each DMC member is a knowledgeable expert, able to understand the study's objectives, results, and impact on patients. The DMC statistician evaluates and interprets the accumulated study results on behalf of the DMC and is a voting member.

Because Tom is the biostatistics team member on a blinded study, he cannot participate directly on the DMC for his study. He works with his colleagues in the biostatistics department to designate a third biostatistician as a liaison between Tom, who must remain blinded, and the DMC statistician, who is often unblinded.

The DMC charter defines the role of the liaison biostatistician. Although liaison biostatisticians operate independently from the study teams, they are usually employed by the sponsor (or by the CRO, if biostatistics services have been outsourced). Tom and his colleagues in the biostatistics department collaborate to designate the liaison biostatistician. Although a biostatistics colleague, who is otherwise not involved with Nancy's study, serves as the liaison biostatistician for her study's DMC, Tom serves as the liaison biostatistician for some of his colleagues' studies.

The liaison biostatistician on Tom's study works with Maria, the data manager, to extract the appropriate subsets of data from the database, obtains the treatment codes from IVRS, and gets the programmed tables and listings code from Tom. Before each DMC meeting, the liaison biostatistician analyzes, summarizes, and submits the accumulated safety data to the DMC for review. During closed DMC meetings, the liaison biostatistician presents and discusses the safety data. Following each DMC meeting, the liaison biostatistician reviews the DMC meeting minutes for accuracy and conscientiously destroys or archives the confidential meeting materials according to the DMC charter's rules.

Interim Analysis

Because Dr. Abernathy's senior managers wanted information about the CanDo Pharma drug's actions as soon as possible, the protocol specified an interim analysis after half of the patients were treated. In the SAP, Tom specified a few key data points for the interim analysis; for efficacy interim analyses, biostatisticians typically analyze only the primary end point data. To conduct this analysis, the statistician must know which treatment (drug or placebo) each patient received. Therefore, Tom asks the liaison biostatistician to perform the interim analysis, so that he can remain blinded.

Most importantly, Tom's SAP specifies how the data and interim analysis results will be sequestered. Tom, Dr. Abernathy, and the other blinded personnel must not have access to these results until after the end of the study; otherwise, they might be tempted to make decisions that influence how the remainder of the study is conducted and how the final data are analyzed. Those decisions would be biased and invalidate the study. The SAP and CanDo Pharma's SOPs specify the uses of the results from the interim analysis and the procedures for communicating those results only to certain designated people.

Tom's most time-consuming work comes at the end of the study. Maria provides him with the final, locked database, and Tom uses the SAP to guide his analysis of the unblinded data.

Data Analysis

Tom, like many biostatisticians, writes his own statistical programs for the data analysis. CanDo Pharma also employs full-time statistical programmers; they assist biostatisticians who are less skilled in creating the customized analysis programs needed for each study. SAS, the industry standard for statistical software, is a validated system accepted by most regulatory agencies. Using his customized SAS programs, Tom populates his programmed tables, graphs, and listings with the corresponding efficacy and safety data points.

ICH regulations require Tom to conduct an extensive analysis of the drug's therapeutic effects. He statistically compares the data points for drug and placebo treatment groups to assess the drug's therapeutic effectiveness. He also analyzes whether other drugs taken by the patients and other medical conditions altered the magnitude or duration of the therapeutic response. He then examines subsets of the patients grouped by age, gender, disease severity, and other factors to determine whether those conditions affect the therapeutic response. Finally, in addition to the grouped data, he generates extensive tables showing each patient's individual response to the drug.

The ICH requirements for reporting safety-related data are even more extensive. Tom classifies the adverse events using the Medical Dictionary for Regulatory Activities (MedDRA) coding system and groups the data by body system (e.g., cardiovascular, pulmonary, renal, etc.) and severity. He uses the methods specified in his SAP to compare the treatment groups for differences in the frequency of adverse event (AE) occurrence, time of onset, and relation to dose level. In addition to the grouped data, he generates detailed tables showing each patient's individual AEs, abnormal laboratory measurements, and the circumstances surrounding each of those adverse responses (e.g., time of onset, investigator site, actions taken to treat the AE, and whether the patient received the experimental drug or placebo).

Tom gives special consideration to analyzing serious adverse events (SAEs). He extracts from the database all of the details surrounding each SAE, such as abnormal laboratory results, vital signs, and clinical observations. In addition to his analytical assessment, Tom gives his compiled SAE tables and listings to Dr. Abernathy or a nurse in the clinical safety group. They are required to write patient narratives, which explain each SAE case in detail from a medical perspective. (See Chapter 10 for more details on patient narratives.)

ICH requires data of sufficient detail in the study report so that statisticians at the regulatory agencies can conduct their own, independent analysis of the drug's beneficial and potentially harmful effects. Because these tables and patient listings are extensive, they will be attached as appendices to the final study report.

In addition, Tom extracts summary tables and graphs from his analyzed data to highlight key findings of the study. Mike, the medical writer, will incorporate these summary tables and graphs into the body of the study report when he writes the text of the study results and conclusions. (See Chapter 10 for details on writing study reports.) Tom writes the sections of the study report that describe the statistical methods and sample size rationale. He takes most of the wording directly from the SAP. If the team made any changes to the study or planned analyses, Tom explains the reasons for the changes and the implications on interpreting the results.

Tom's final responsibility on Nancy's study is to archive his statistical programs, the SAP, and all related documents so that they are available to internal auditors and regulatory inspectors, if requested.

In addition to his work as a clinical study biostatistician, Tom and his colleagues also assist the laboratory scientists at CanDo Pharma in analyzing their data. Unlike clinical teams, the scientists who work in research laboratories have much greater freedom to explore novel and innovative scientific ideas without regulatory constraints. Their studies are often spontaneous, and they freely adjust their procedures while the studies are ongoing.

Research scientists usually consult Tom only after the end of their experiments and ask for his assistance. Therefore, his first job is to put the data in proper analytical order. This can be challenging, because the studies may have many uncontrolled variables, unequal group sizes, and a small number of experimental observations. He may spend considerable time organizing the data and preparing a suitable database before he can begin his analysis.

The flexible design and uncontrolled variables may also require Tom to use more creative statistical methods to analyze these data. However, exploratory study data provide an opportunity for Tom to sharpen his skills as a biostatistician and perhaps devise a novel statistical approach. Such opportunities are rare with clinical data analysis, which is much more closely regulated and must meet very high standards for data integrity.

ONE OF TOM'S DAYS

When Tom arrives at his desk, he hopes to spend the day working on two tasks: conducting the final analysis of the data from Nancy's clinical study and attending a meeting that Dr. Abernathy has scheduled to discuss plans for a new clinical study of the CanDo Pharma drug.

Yesterday, Maria, the data manager, officially notified Nancy's study team that she had locked the database, and the blinded treatment codes have been broken. Tom is now able to proceed with his final analysis of the unblinded data. Because the study has taken several years to complete, Nancy's clinical study team and Dr. Abernathy's strategy team are anxious to learn the results. The complete analysis will be long and laborious, but Tom will prioritize his work and analyze the most important datasets first. This "preliminary analysis" will focus on the extent of the drug's primary therapeutic effects and most prominent adverse effects.

Tom and Dr. Abernathy hope that the results verify the study's objective, which was to show that the CanDo Pharma drug worked better than the placebo treatment. If the preliminary analysis confirms this outcome, the strategy team will proceed with their original plans for the next clinical studies of the CanDo Pharma drug. However, if the analysis reveals a less positive outcome, Dr. Abernathy's team may need to wait until Tom completes the rest of his analysis to learn more about the drug's properties and limitations.

To complete his full analysis, Tom will follow his statistical analysis plan and systematically analyze all of the specified study data, including a detailed examination of various data subsets. He may find, for example, that patients over 65 years old respond less well than younger patients.

Dr. Abernathy's team might then choose to exclude elderly patients from the next clinical study or give them a higher dose of the drug.

Tom is aware of all of these possible complications as he starts his preliminary analysis. Even if things go smoothly, he will need several weeks to complete the full analysis. However, this morning he only has time to begin organizing his analysis activities before leaving his desk to attend Dr. Abernathy's meeting.

Dr. Abernathy has invited Tom and a designated study manager to join the strategy team for their discussion on the design of a new clinical study of the CanDo Pharma drug. Until Tom provides the results of his preliminary analysis, the team will assume that Nancy's study was successful in showing beneficial therapeutic effects. The new study will use an improved formulation of the drug, and Dr. Abernathy wants to know how many patients they need to include in the study to obtain definitive results. The team prefers to include fewer patients, so that the study will finish more quickly and be less expensive. However, Tom recommends a larger patient sample, based on statistical consideration of the known study variables.

This new study will include not only a placebo group but also a group of patients who will take a standard, commercially available drug treatment. The team discusses the choices in study design for comparing the effects of the CanDo Pharma drug to the competitor drug. Should it be a superiority, noninferiority, or equivalence study? Superiority studies are designed to demonstrate that one treatment is better than another; noninferiority studies demonstrate that a treatment is not worse than another; and equivalence studies demonstrate that one treatment is as effective as another.

Each study design has advantages and disadvantages. The marketing representative on the team wants a superiority study, because the data will allow the company to make strong advertising claims that the CanDo Pharma drug is better. However, collecting data that demonstrate superiority requires a larger patient sample and careful patient selection criteria. Even so, Tom knows that superiority is difficult to demonstrate statistically. Noninferiority and equivalence are lower statistical standards and more easily achievable, considering the variability in patient responses to most drugs. At the end of the meeting, Dr. Abernathy suggests that they wait until Tom finishes his preliminary analysis of Nancy's study. Those results may give them more confidence in choosing the most appropriate design for the new study.

When Tom returns to his desk, he feels even greater urgency to finish his preliminary analysis quickly, and he gets right to work. Tom is doing his own statistical programming for this study, and he spends the next few hours reviewing the programs he has written for the preliminary analysis. By mid-afternoon, he has finished conducting tests with artificial data to confirm that the programs are running properly.

Just before he starts his preliminary analysis of the study data, one of his biostatistics colleagues stops by to ask a favor. The colleague, who is the liaison biostatistician for a study in another CanDo Pharma development program, has been preparing for a DMC meeting, which will be held tomorrow. He explains that his whole family has come down with the flu, and he is now feeling symptoms, too. He is going home early and doubts that he will be able to work tomorrow. He asks if Tom would take his place at the DMC meeting. Tom is one of the few biostatisticians who is not involved with this clinical study and therefore one of the few people in the department eligible to substitute for his colleague.

Although he would prefer to continue analyzing the data from Nancy's study, Tom agrees to help. Fortunately, his colleague has finished all of the statistical preparations for the DMC meeting. But because Tom is unfamiliar with the study, they spend the next hour going over the details.

The study is well underway. Most of the patients have already enrolled, and about half of them have completed their scheduled treatments. Under its charter, this DMC's responsibilities are limited to assessment of the study's safety data, but the study has accumulated a lot of data points. The liaison biostatistician has downloaded the AE and SAE datasets, organized them by treatment group (i.e., A, B, C, etc.), generated patient listings, and prepared summary tables. His analysis indicates that there is no difference in side effects between the groups, and the frequency and severity of adverse reactions are within the range reported at earlier DMC meetings. Therefore, he expects that the DMC's discussion about the data will be routine and that no further analysis will be requested.

Tom asks a few questions to clarify details of the study and the analysis procedures that his colleague used. As his colleague leaves, he asks Tom to take good notes and promises to do all of the follow up tasks when he returns to work. He also promises to stay near his home phone, if his input is needed during tomorrow's meeting.

Tom is confident that he can handle the DMC presentation, but he spends the rest of the afternoon going over his colleague's notes and the previous DMC meeting minutes. He will also plan to arrive at work early tomorrow and review the data one more time before the DMC meeting starts. Hopefully, the DMC will not ask him to do any immediate follow-up work, and he can then return to his desk to continue working on the preliminary analysis of Nancy's study.

HOW TOM GOT HIS BIOSTATISTICS JOB

Requirements

Tom always enjoyed studying mathematics and using computers. He took an advanced placement statistics course in high school and decided to major in mathematics in college. He also took a number of science courses. As he learned more about science and statistics, Tom realized that he wanted to pursue a career in applied statistics. Unlike theoretical statistics, which focuses on the theorems and proofs of classical statistics concepts, applied statistics uses statistical methods to analyze real data sets that are collected during scientific investigations. After graduating, Tom continued his statistics studies and earned a master's degree in biostatistics.

BOX 6-2. Requirements for a Biostatistics Position

- Master's or Ph.D. degree in biostatistics or computer science
- Knowledge of advanced statistical methodology
- Experience with statistical software such as SAS and S-plus
- Ability to communicate statistical information to nonstatisticians
- Knowledge of ICH requirements for statistical analysis
- Ability to collaborate and promote team goals
- Excellent written and oral communication skills
- Ability to work under stress and tight deadlines
- Excellent attention to and accuracy with details
- Knowledge of clinical data management principles
- Statistical programming experience is helpful
- Experience with CDISC standards is helpful

Biostatisticians always work with data collected by someone else. Tom was fortunate that his graduate school department frequently collaborated with the university's medical center on clinical research projects. One of the medical center physicians needed assistance with analysis of the data from his independent clinical investigations, and this collaboration between Tom and the physician was mutually beneficial. Tom used the clinical data as the basis of his master's thesis, and the physician incorporated Tom's analysis in a research paper, which was eventually published in a medical journal.

Many universities and schools of public health offer master's degree and Ph.D. programs in biostatistics. Faculty members who advise students in advanced degree programs select data for biostatistics analysis based on their individual academic interests. In addition to medical studies, these interests might include data from environmental, epidemiology, or marine biology studies. Alternatively, some biostatisticians earn a graduate degree in computer science and then guide their career toward biostatistics or statistical programming.

Tom's faculty advisor in graduate school encouraged him to join a statistics professional society. Three prominent societies for biostatisticians are the American Statistical Association (ASA), Statisticians in the Pharmaceutical Industry (PSI), and the International Biometric Society (IBS). PSI covers biostatistics specific to the pharmaceutical industry. ASA covers all areas of statistics but has a large and active section devoted to biopharmaceutical statistics. IBS covers biostatistics exclusively. These organizations sponsor journals, conferences, and sub-specialty training on pharmaceutical statistics. Through their parent organizations and local chapters, they also offer opportunities for networking, career development, and job postings.

Tom's advisor also encouraged him to sharpen his skills in statistical programming using SAS, the software standard accepted by most regulatory agencies. He took advantage of an online training program and earned certification as a SAS programmer. The training gave Tom a better understanding of how to organize complex data sets and customize data analyses. SAS certification also gave him a significant advantage when applying for biostatistics positions.

Through the physician who provided the data for his master's thesis, Tom met several CRAs who were monitoring industry-sponsored clinical studies at the medical center. They encouraged Tom to apply for a biostatistics internship. Several large pharmaceutical companies offer summer internships to students enrolled in master's and Ph.D. degree programs. Interns work under the direction of senior biostatisticians and learn clinical study design and data analysis methods used in drug development. The ASA posts information about biostatistics internships on its website.

As Tom learned more about analyzing the clinical data for his thesis, he realized that he wanted to pursue a career in medical biostatistics. He liked applying theoretical statistics concepts to real medical situations. He liked adapting the textbook criteria for statistical design and analysis to complex experimental studies. And he liked the challenges in extrapolating the results of a small sample to make predictions about the effects on an entire patient population.

Finding a Biostatistics Position

You can find entry-level biostatistics positions at large sponsor companies and at large CROs. These companies, along with ASA, PSI, and IBS, post job openings on their websites, which can easily be searched by job title and category. The job titles for entry-level biostatistics positions vary considerably between companies, but look for titles such as biostatistician, biostatistics analyst, or biostatistics programmer.

Large companies offer better opportunities for an entry-level biostatistician than small companies. Large sponsor and CRO companies typically have fully developed biostatistics departments. The experienced staff and established procedures offer the best opportunities for high-quality training, experience, and career advancement.

Small companies rarely hire entry-level biostatisticians. Small CROs have limited resources and typically hire only experienced statisticians, who are expected to implement all biostatistics services requested by the CRO's clients. Small sponsor companies also have limited resources and typically outsource most of their statistical analysis work to a large CRO or a CRO that specializes in statistics services. In both cases, the biostatisticians at small companies must be experienced and highly qualified, because they are responsible for the quality and appropriateness of the statistical analyses, whether they conduct the work themselves or oversee the work of contractors.

Hiring managers at both CROs and sponsor companies look for the same skills and qualities in their job candidates: statistics training, statistical programming experience, and computer database familiarity. If you have a master's or Ph.D. degree in biostatistics, applied mathematics, or computer science, you qualify for entry-level positions.

Hiring managers expect that most entry-level statisticians do not have previous experience with clinical studies, even if the candidates have analyzed or programmed medical data. Therefore, they look for work habits, accomplishments, and other skills that are characteristic of successful industry biostatisticians. (See Box 1-2 in Chapter 1 for these general requirements.) They prefer candidates whose thesis projects included analysis of biological (preferably human) data. They also look for candidates who have successfully collaborated with scientists or medical specialists.

Landing the Job

All of the hiring managers ranked Tom as a highly desirable biostatistics candidate. Although he had never analyzed data from an industry-sponsored clinical study, Tom's collaboration with a medical center physician, the data in his master's thesis, and his internship at a pharmaceutical company gave him clinically relevant experience. Furthermore, his SAS certification confirmed his skills as a SAS programmer.

BOX 6-3. *Tips for Getting a Biostatistics Position*

- Gain experience analyzing biological or clinical data
- Gain experience presenting, defending, and discussing statistics concepts and results
- Take online courses on Good Clinical Practices
- Gain experience with medical terminology and clinical concepts
- Consider volunteering at a medical center to assist with data analysis
- Consider obtaining SAS programming certification
- Consider becoming familiar with CDISC standards
- Join a professional society such as ASA, IBS, PSI, or PharmaSUG
- Search biostatistics job postings on websites of drug, medical device, and CRO companies
- Preferentially look for entry-level jobs at large CROs and biopharmaceutical companies

Tom's experiences gave him the necessary qualifications for an entry-level biostatistics position, but there are other ways. Online courses on topics such as good clinical practices (GCPs), drug development activities, MedDRA coding, and Clinical Data Interchange Standards Consortium (CDISC) conventions enhance a biostatistician's understanding of clinical study and clinical data handling concepts. Also, scientists and physicians at large medical centers and universities often seek advice from statisticians. Student biostatisticians who assist those researchers sometimes become coauthors on published scientific papers.

The hiring managers were less impressed with the candidates who competed with Tom for entry-level positions, even though some had previous experience with clinical studies. Study teams expect the biostatistician to have excellent communications skills and the ability to propose study designs and statistical analysis methods in a clear, understandable manner. Therefore, most hiring managers invite candidates to make a formal statistics presentation as part of the interview process. They reject candidates who are unable to explain and justify their statistical methods in a formal presentation. They also reject candidates who are unable to evaluate and discuss alternative analysis strategies proposed by the audience.

Working for a CRO Versus Sponsor Company

Biostatisticians at CROs often work for clients that are small, start-up companies, because those sponsors have no internal biostatistics expertise. When CRO-based biostatisticians work for large client companies, they must ensure that their statistical approach is consistent and compatible with the analysis procedures used for other studies in the client's development program. Through their various work assignments, therefore, biostatisticians at CROs quickly gain broad experience across many protocols as well as across the standards and cultures of diverse clients. For this reason, hiring managers at sponsor companies often consider biostatisticians with a CRO background to have a significant advantage.

However, it is also beneficial for CRO biostatisticians to have previous experience working for one or more sponsor companies. Such experience provides a greater understanding of the needs of the sponsor and the analysis and reporting processes that sponsors use. Biostatisticians at sponsor companies gain a deep understanding of the regulatory require-ments, operational systems, and business drivers involved with analyzing data from drug and medical device studies—an insight that is otherwise difficult for CRO-based biostatisticians to appreciate. However, the purely statistical aspects of the work are the same at both CRO and sponsor companies.

Although Tom could have taken an entry-level biostatistics job at a large CRO, he decided to accept the offer from CanDo Pharma as a SAS programmer. Under the direction of experienced biostatisticians, Tom wrote statistical analysis programs for several clinical studies. In parallel, he became familiar with CanDo Pharma's SOPs and attended courses on drug development, ICH guidelines, and GCPs offered by CanDo Pharma. After Tom mastered the regulatory requirements, clinical con-cepts, and SOPs for statistics, quality assurance (QA), and quality control (QC), he took greater responsibilities as a biostatistician.

By the time Nancy's study started, Tom had been promoted to biosta-tistical analyst and had served as the study biostatistician on several other clinical studies in Dr. Abernathy's development program on the CanDo Pharma drug. He was fully prepared to conduct all the biostatistics activ-ities as a member of Nancy's clinical study team.

MOVING FORWARD

Tom decided to continue his career in biostatistics and gain more expe-rience designing clinical studies and analyzing clinical data at CanDo Pharma. Although the titles of biostatistics positions (e.g., senior biosta-tistician, principal biostatistician, manager of biostatistics) vary consid-erably, large sponsor companies and large CROs offer career tracks in biostatistics. From an entry-level position, biostatisticians can expect promotions every 2–4 years to the next level of responsibility, depending on performance. Typically, biostatisticians like Tom reach a manager-level position after gaining experience at two or three lower levels.

Biostatistics managers supervise junior-level statisticians and take greater responsibility for handling clinical data. They may also serve as a DMC liaison biostatistician, draft DMC charters, and recommend DMC members. Experienced biostatistics managers may serve on strategy teams, which steer the activities of entire development programs.

Alternatively, Tom could choose to specialize in a subdivision of biostatistics work at either a sponsor company or CRO. Some biostatisticians support the development program by pooling data from multiple clinical studies and preparing two important sections of the Marketing Authorization Application (MAA)—the summary of clinical efficacy and the summary of clinical safety.

In the summary of clinical efficacy, biostatisticians compile the data from all of the clinical studies that examined the therapeutic effects of an experimental drug. Even though the data of individual studies were previously analyzed, the biostatistician must merge the data, conduct a comprehensive analysis of the results across all of the studies, and draw overall conclusions about the drug's therapeutic benefits.

Similarly, in the summary of clinical safety, biostatisticians compile the AE and SAE data that were reported in all of the clinical studies using an experimental drug. Even though the number of AEs and SAEs might be small in individual studies, the biostatistician must analyze the accumulated side effects across all of the studies to determine the drug's overall safety profile. (See Chapter 8 for details of MAA preparation.)

Other biostatistics specialties in CROs and biopharmaceutical companies include supporting early stage clinical studies or preclinical studies. Statistical analysis for Phase 1 clinical studies emphasizes assessment of treatment side effects and the time course of drug concentrations in the blood. Preclinical statistics focus on analysis of toxicity results from animal studies and complies with statistical requirements in the good laboratory practices (GLP) sections of the ICH standards.

Most biostatisticians continue to develop their technical skills in programming and their biostatistics expertise, rather than move into another clinical career track. Biostatisticians with good management skills can advance to positions of greater responsibility and eventually head a biostatistics or biometrics department. Their promotion to this level is based not only on their technical expertise but also their ability to think strategically and interpersonal skills such as demonstrating good judgment, inspiring quality performance in others, and working collaboratively.

Alternatively, some experienced biostatisticians work independently as consultants and contract their services to sponsor companies and CROs. Most of their clients are small, start-up companies that do not have internal biostatistics resources. Biostatistics consultants can charge higher fees than employer-based biostatisticians and have the flexibility of accepting or refusing specific work assignments. However, they depend on their reputation for obtaining new clients and retaining existing clients. So, they work especially hard to ensure the quality of their work, take the initiative to learn about new and changing regulatory requirements, and actively participate in biostatistics professional societies.

BIOSTATISTICS RESOURCES

Good Clinical Practice Training and Certificate Programs

ClinfoSource (www.clinfosource.com) offers online GCP courses and certification for CME and CNE credit at reasonable cost.

Kriger Research Center International (www.krigerinternational.com) offers online GCP courses and certification in multiple languages on the country-specific web pages under the "Courses" tab.

Association of Clinical Research Professionals (www.acrpnet.org) offers online GCP courses and certification for CME, CNE, and ACRP credit at reasonable cost.

GCP Training Online (www.gcptraining.org.uk) offers online GCP training and certification, emphasizing the European Clinical Trial Directive, at reasonable cost.

Statistical Programming and Coding Conventions (Including Certifications)

SAS programming (www.support.sas.com/certify) offers online courses and a global certification program in SAS programming.

Kriger Research Center International (www.krigerinternational.com) offers online statistical software programming course including an introduction to SQL and SAS. Course and diploma certification information are in multiple languages on the country-specific web pages under the "Courses" tab.

Medical Dictionary for Regulatory Activities (www.meddramsso.com) is a clinically validated dictionary of terms used to report adverse event data from clinical studies.

Clinical Data Interchange Standards Consortium (www.cdisc.org) sets clinical data conventions for acquiring, exchanging, submitting, and archiving clinical data.

Medical Dictionaries and Terminology

ClinfoSource (www.clinfosource.com) offers an online medical terminology course for CME and CNE credit at reasonable cost.

Kriger Research Center International (www.krigerinternational.com) offers an online course in medical terminology in multiple languages on the country-specific web pages under the "Courses" tab.

Free-ed.net (www.free-ed.net/free-ed/healthcare/medterm-v02.asp) offers an online course in medical terminology and CEU credit at reasonable cost.

Des Moines University (www.dmu.edu/medterms) offers a public access, online course in medical terminology.

Biostatistics Internship Programs

ASA Center for Statistics Education (www.amstat.org/education) posts information about biostatistics internships under the "Graduate/Internships" tab on the Education web page.

Regulatory Standards and Guidelines

International Conference on Harmonization (www.ich.org) sets international standards for clinical studies of investigational drugs.

International Organization of Standardization (www.iso.org) sets international quality standards recognized by governments for commercial products.

European Medicines Agency (www.ema.europa.eu) evaluates and supervises safety regulations of drugs in the European Union.

European Commission/medical devices (http://ec.europa.eu/enterprise/sectors/medical-devices/index_en.htm) sets regulatory standards for medical devices in the European Union.

Food and Drug Administration (www.fda.gov) sets safety regulations for foods, drugs, and other medical products in the United States.

Biostatistics Professional Organizations

American Statistical Association (www.amstat.org) promotes educational services, proper application of statistics, and opportunities for advancement in statistics to statisticians and users of statistics.

Statisticians in the Pharmaceutical Industry (www.psiweb.org) provides a forum for regular discussion of statistics issues relating to the practice of statistics in the pharmaceutical industry and promotes good statistical practice within the industry.

International Biometric Society (www.tibs.org) is an international society promoting the development and application of statistical theory and methods in the biosciences.

Pharmaceutical Industry SAS Users Group (www.pharmasug.org) is a forum for the exchange and promotion of new ideas concerning the use of SAS software as it relates to quantitative health sciences, including the pharmaceutical and medical device industry.

Profiles of Sponsor and CRO Companies (*See detailed list in Chapter 13*)

Pharmaceutical Research and Manufacturers of America (www.phrma.org) represents the leading pharmaceutical research and biotechnology companies in the United States.

Biotechnology Industry Organization (www.bio.org), the world's largest biotechnology organization, represents more than 1200 biotechnology companies.

Medical Design & Manufacturing (www.devicelink.com) is an online resource for the medical device industry.

Association of Clinical Research Organizations (www.acrohealth.org) represents the world's leading clinical research organizations.

Salary Surveys for Biostatisticians

American Statistical Association (www.amstat.org) is published annually. Access the most recent survey by entering "salary survey" in the website's Search field.

Medical Device & Diagnostic Industry Salary Survey (www.devicelink.com) is published periodically. Access the most recent survey by entering "salary survey" in the website's Search field and selecting the Research and Development survey.

Salary.com (www.salary.com) publishes salary ranges for positions in a wide range of industries. Enter "biostatistician" and if appropriate your targeted zip code in the Salary Wizard.

7

Entering as a Clinical Quality Assurance Auditor

CLINICAL QUALITY ASSURANCE (CQA) auditors inspect clinical study documents, clinical department processes, investigator sites, and contract research organizations (CROs), all of which must comply with Good Clinical Practice (GCP) regulations, standards, and guidelines and the sponsor's standard operating procedures (SOPs). A growing number of countries and a proliferation of International Conference on Harmonization (ICH) and International Organization for Standardization (ISO) regulations require sponsors and CROs to comply with increasingly specific regulatory requirements. So, in today's global environment, compliance is quite a challenge for sponsors, who must observe the most current rules, and for CQA auditors, who must ensure that they do.

The CQA department is organizationally separate from the clinical department and therefore can independently oversee the activities of Nancy's study team and Dr. Abernathy's strategy team. Most sponsors, like CanDo Pharma, establish their own CQA departments, which typically report directly to the sponsor's senior management. Many CROs also maintain a CQA function, both to manage the quality of their employees' work and to offer CQA services to their clients.

THE ROLE OF THE CLINICAL QUALITY ASSURANCE AUDITOR

Clinical study teams generally do not include a CQA representative. The team manages the day-to-day quality of its studies by following the requirements of regulatory agencies, the company's SOPs, and individual job descriptions. However, many sponsor companies periodically monitor the quality of clinical operations using internal auditors from their CQA

BOX 7-1. CQA Auditor Responsibilities

- Conduct internal audits of clinical studies
- Conduct audits of clinical investigator sites
- Conduct audits of contracted vendors and CROs
- Ensure compliance with ICH, ISO, FDA, and other regulatory requirements
- Ensure compliance with sponsor and vendor SOPs
- Write audit reports of audit findings
- Collect responses to audit findings and documents corrective actions
- Host and cooperate with inspections by regulatory authorities
- Serve as compliance consultant to clinical teams
- Manage SOPs
- Train clinical teams on quality assurance topics

departments or through the use of CRO auditors engaged by the clinical department.

The relationships and interactions of the CQA unit are illustrated in Figure 7-1. CQA managers randomly select a portion of clinical studies to audit, based on the nature of the study, the study's documented risk,

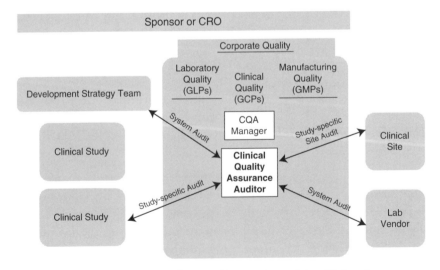

Figure 7-1. Relationships and interactions of the CQA unit.

and the product's classification. (The Food and Drug Administration [FDA], for example, uses four product classifications for drugs and three for medical devices.) Because different countries use different classifications and considerations for risks, especially for medical devices, auditors must keep these various perspectives in mind when they conduct their audits.

Routine CQA Study Audits

Typically, CQA managers select 5%–10% of ongoing and completed clinical studies to audit annually. In addition, many CQA departments require that every final protocol be audited to ensure compliance before the study begins and that all final study reports be audited to ensure that they accurately reflect the documented evidence stored in the study files.

Nancy's clinical study was one the CQA managers selected to audit, and they assigned Carlo as the auditor. Carlo does not have supervisory responsibilities, but the CQA department provides a few administrative assistants, who assist the CQA auditors with clerical and administrative tasks. Carlo's purchasing authority is limited to small office supply items, but the CQA department incorporates the cost of auditing activities in its annual budget. Because Carlo's auditing assignments often involve travel to investigator sites or external vendor facilities, CanDo Pharma reimburses his travel expenses, as long as he follows the company's travel expense guidelines.

Carlo began his audit of Nancy's study after most of the patients had been enrolled but while they were still being treated. The advantage of auditing an active study is that the team can use the audit findings to make immediate quality improvements that will benefit the study. However, Carlo also conducts some audits after the studies are completed. Although the audit findings will not benefit a closed study, the clinical teams can make adjustments to improve the quality of their future studies.

Assessing Regulatory Compliance

CQA auditors like Carlo assess the quality of a clinical study by determining whether the study team is following established quality standards. Quality standards are specified in laws, rules, regulations, and guidelines issued by official bodies such as the ICH, the ISO, the FDA, and the European Medicines Agency (EMEA).

GCPs issued by the ICH are a key guideline for clinical study conduct. The regulatory authorities in many countries elaborate on the ICH guidelines and require sponsors to meet additional quality standards. In the United States, for example, these requirements are published in the Code of Federal Regulations (CFR), Title 21, which articulates specific requirements for communicating clinical study information to the FDA. In Europe, the Clinical Trials Directive, issued by the European Parliament, specifies the requirements for conducting and reporting clinical studies, but defers to individual European Union (EU) member countries for submission and regulatory review procedures.

In addition to GCPs and standards required by regulatory agencies, sponsors like CanDo Pharma establish SOPs, which state how they will implement the regulatory requirements at their companies. Each SOP includes the step-by-step procedure for completing a required task and identifies who (by job title) is responsible for performing each step. CanDo Pharma's SOPs for clinical studies comply with all regulatory requirements and ensure that tasks are conducted the same way on every study, even when substitute members join the teams.

One of the best ways for Carlo to determine whether Nancy's study team is following all the appropriate regulatory requirements, GCPs, and SOPs is to examine the study documents in the team's master file and at the investigator sites. For Nancy's study, which is being conducted at many clinical sites in three countries, Carlo will visit a portion of the sites—typically, ones that have enrolled large numbers of patients.

Properly signed and dated documents (or their electronic equivalents) should provide an accurate and complete record of all activities and events that occur during the study. If an event or task is not documented, the auditor assumes it did not happen. Sometimes, Carlo also reviews source documents. Although these are a more informal series of notes and comments generated during the clinical trial, they often help him to understand the sequence and significance of study activities. Carlo is particularly interested in documents that verify activities in four categories: protecting patients' rights, reporting adverse events, following the study protocol, and handling the data.

Ensuring Patient Welfare. Carlo knows that regulatory authorities expect CanDo Pharma's teams to protect patients' rights and welfare. The

Institutional Review Board (IRB) approval letter confirms that the IRB reviewed the protocol and informed consent form (ICF), found no concerns regarding patient safety or welfare, and authorized the principal investigator (PI) to conduct the study. The patient's signature and date on the ICF confirms that, before the start of the study, the PI (or designee) explained all study procedures, identified the risks, and satisfactorily answered all of the patient's questions, and that the patient voluntarily agreed to participate. Carlo checks the study binders to make sure the study coordinator filed these documents properly.

Regulatory authorities also expect all study workers to comply with requirements for adverse event (AE) and serious adverse event (SAE) reporting. During his audit, Carlo carefully reviews all of the documents related to patient safety. He compares safety reports with the original case report forms (CRFs) and patient charts to confirm that the AE and SAE information submitted to regulatory authorities was accurate, complete, and submitted on time. In addition, he searches the patient charts and CRFs for any adverse events that the PIs should have reported to the study team but did not.

Protocol Compliance and Data Handling. Carlo determines whether the study team and PIs are following regulatory requirements and SOPs, as they relate to the protocol, by examining the documents in the master study file at CanDo Pharma and the study binders at the investigator sites. He confirms that the study personnel properly signed, dated, and filed regulatory documents such as the study protocol, investigator's brochure, laboratory certifications, study correspondence, patient screening logs, the roster of study personnel and their assigned responsibilities, the IRB membership list, and the Data Monitoring Committee (DMC) membership list. The study team must also file signed and dated contracts for all investigator sites and vendors (such as specialty laboratories, consultants, and CROs).

Carlo also reviews documents associated with drug handling. Regulatory agencies and CanDo Pharma's SOPs require the study staff to keep all experimental drugs and placebos safe and secure. Carlo checks the drug accountability forms and shipping documents to confirm that these supplies were shipped, stored, dispensed, and returned using the appropriate procedures.

A similar approach is used for auditing medical device accountability. Of particular interest is the disposition of medical devices that are returned at the end of the study. Sometimes, those devices are destroyed, but often they are returned to the sponsor's laboratories for researching design changes or use in another project. Therefore, auditors carefully review the sponsor's process for receipt, accountability, and disposition of medical devices when they are returned.

Data verification and validation is one of Carlo's most time-consuming tasks. He compares the data in the clinical database with the original entries on the CRFs and related queries. Because Nancy's study uses electronic data capture (EDC), the database contains an audit trail of each entry and change to the electronic CRFs and database, including electronic signatures and date stamps. Carlo checks the dates and signatures to confirm that only authorized people have accessed the database. He also checks the queries to confirm that they were handled and resolved properly.

Debriefing

At the end of his visit to each investigator site, Carlo meets with the PI and other key site personnel and gives them a brief summary of his observations at the site. If he found deficiencies or discrepancies during his audit, he may probe deeper and ask the site staff to clarify the issue. Although this is an informal and unofficial meeting, the site staff appreciates getting immediate feedback from Carlo and the chance to discuss any areas of concern before he leaves.

When Carlo has completed his audit of the selected clinical sites and the study master file, he schedules a debriefing meeting with Nancy's team and key managers at CanDo Pharma. Carlo presents his main findings and allows the team to provide input before he issues his official audit report. These are valuable discussions because clinical studies often involve unusual procedures or specialized equipment, whereas the published regulations only discuss clinical procedures in general terms. Carlo must therefore interpret the intent of the regulatory requirements as they apply to Nancy's study and then decide whether the team has complied. If his interpretation is faulty because he lacks a complete understanding of the study's technical or scientific details, he may have cited a regulatory violation inappropriately. At the debriefing meeting, Nancy's team has the opportunity to clarify these misunderstandings with Carlo before he writes his report.

Audit Report

Using his audit notes and the team's input, Carlo then drafts his audit report. Although he is careful to compliment the team on activities they performed correctly, he writes detailed comments on the deficiencies and discrepancies he observed. In each category of his comments (such as handling AEs, pharmacy activities, and monitoring), he gives specific examples of documents or procedures that were deficient and cites the relevant published regulations. Carlo also indicates the severity of each finding based on the definitions in CanDo Pharma's SOPs. Typically, these are categorized as "critical," "major," and "minor," and the designation determines the importance and urgency of the corrective actions. In a separate document that accompanies the audit report, Carlo suggests options for addressing the critical findings. Although Nancy and the team are not obligated to follow his suggestions, they find his input helpful and constructive.

The CQA department formally issues Carlo's written report to Nancy's team, along with a deadline for responding to his audit findings. Audit findings that affect patient welfare or the integrity of the study data are usually considered critical, and the team must take corrective actions to minimize the chance of repeating those problems. Each corrective action depends on the nature of the infraction: The team may redesign the study forms to facilitate correct data entry, modify the steps or responsibilities in an SOP to eliminate ambiguities, or schedule supplemental training to ensure proper implementation of study procedures.

Carlo's department will keep the audit "open" until he receives the team's responses to each key finding. An acceptable response may be either a plan for taking corrective actions or confirmation that the team has completed the corrections. When Carlo receives the team's responses, he will review the actions proposed or taken and determine whether they address the findings adequately. If so, he formally closes the audit and archives his audit report. If the responses are not satisfactory, Carlo works with Nancy to clarify or resolve the outstanding issues. After the issues are reconciled, Nancy updates her responses and Carlo then closes the audit.

Internal Systems Audits

In addition to study-specific audits, Carlo also conducts systems audits. Systems audits evaluate the adequacy of an individual clinical process, or system, and usually encompass several studies. Internal processes that are commonly audited include adverse event reporting, site monitoring, data management, and document archiving. Other specialized systems audits include computer validation and external assessments of contract laboratories, CROs, or vendors.

Carlo's department gives priority to clinical systems audits of important internal processes at CanDo Pharma, such as the SAE reporting procedure. Several CQA auditors conduct a systems audit jointly, because it involves many activities. The systems audit of the SAE reporting procedure, for example, includes assessing CanDo Pharma's adverse event SOPs, reviewing safety data, interviewing clinical team members who have safety-related responsibilities, and reviewing SAE reports submitted to regulatory authorities.

Rather than reviewing all of the SAE data on all of CanDo Pharma's clinical studies, Carlo and his coworkers select a small group of studies to audit. Collectively, these studies represent the range and types of SAE data from all of CanDo Pharma's studies—different experimental treatments, different therapeutic areas, and different types of patients.

Through their interviews and document reviews, the auditors assess whether the teams followed the SAE procedures specified in CanDo Pharma's SOPs and the relevant regulatory requirements. In addition, Carlo and the auditors may pick a few SAEs from the selected studies and trace the actions taken by the team from the first observation of the SAE at the clinical site until it was resolved. All of the team's actions should be documented in the study files or in the safety database, as specified by GCP requirements and CanDo Pharma's safety SOPs.

For systems audit reports, Carlo and his coworkers follow the same procedures used for debriefing, drafting, and finalizing study-specific audit reports. The groups responsible for taking corrective actions depend on the nature of the findings, but generally one functional area takes the lead. In the case of an audit of SAE reporting procedures, for example, CanDo Pharma's safety department coordinates the responses and ensures that the appropriate corrective actions are taken, even if some involve other departments.

External Systems Audits

For external audits, Carlo's department gives priority to audits of key contractors and vendors. For example, because CanDo Pharma depends on an external clinical laboratory to analyze the blood samples collected from its clinical studies, the CQA department periodically audits that laboratory. Carlo makes an appointment to visit the laboratory for the audit, which may take several days. At the laboratory, Carlo, along with several other auditors, interviews the staff, reviews the laboratory's SOPs, inspects the facilities, and reviews the laboratory's documents and data storage systems.

The auditors assess whether the clinical laboratory is following its own SOPs, CanDo Pharma's SOPs for sample handling, and the corresponding regulatory requirements. As with internal audits, Carlo and his colleagues conduct an informal meeting with the laboratory management and debrief CanDo Pharma's management. After these meetings, Carlo and his coworkers use their audit notes to draft the audit report. In addition to the main report, they add a supplemental section, which elaborates on their observations for CanDo Pharma's benefit.

Carlo's department formally issues the main audit report to the clinical laboratory, along with a deadline for responding to the audit findings. CanDo Pharma's management team receives the annotated version of the report, including the supplemental section of comments.

The clinical laboratory must respond to the audit report, either by explaining the corrections it has made or presenting a plan for taking corrective actions. While the external audit is "open," Carlo monitors the clinical laboratory's progress in addressing the audit findings. He also keeps CanDo Pharma's executives informed, because some issues stemming from the audit may require negotiation between the laboratory and the decision makers at CanDo Pharma. When Carlo receives responses that are acceptable to CanDo Pharma, he formally closes the audit and archives the audit report.

Special Audits

Although the CQA department maintains an ongoing schedule of internal and external audits, others at CanDo Pharma may request a special clinical audit. For example, a clinical team may request an external audit if they suspect that a vendor is not following appropriate procedures. Rather than conducting its own investigation, the team relies on the CQA auditors to assess the vendor. The clinical department may also request an audit of its internal systems before an inspection by regulatory authorities. Because regulatory agencies can impose fines and take other legal action, sponsors use their own auditors to identify problems and proactively take corrective actions, thus minimizing the chances of negative findings during the regulatory inspection. (See Chapter 8 for more information about regulatory audits.)

ONE OF CARLO'S DAYS

Today, Carlo is returning to his desk at CanDo Pharma after spending several days visiting clinical sites for a routine audit of one of the company's clinical studies. He plans to spend the day working on the first draft of his audit report and simultaneously preparing his slides for the internal debriefing meeting with the study team.

Using his notes and worksheets, Carlo settles at his desk to write his report. A report template that specifies the content and format for all CanDo Pharma audit reports helps him organize his thoughts. He tackles the observations that fall into the critical category first, because those are the ones that potentially affect patient welfare and must be addressed quickly by the study team. The report and his presentation slides will contain a brief description of each critical audit finding and cite the published regulatory standard or requirement that should have been followed.

Carlo also prepares detailed reference notes, in case the team asks for clarification during his presentation. He has interpreted the generalized GCP standards and used his judgment to determine whether those requirements have been violated, and he wants to be prepared to explain and justify his decisions to the study team.

Carlo's draft report and presentation slides also list brief descriptions of his major and minor audit findings. Although the study team must also address these issues, the seriousness and time frame for corrections are less pressing than for the critical findings. Carlo will be prepared to discuss these at the debriefing, but the study team will probably be more concerned about the critical findings.

The CQA department's SOP regarding audits specifies a rapid time frame for issuing a final audit report and the steps that must be completed before it is issued. So, Carlo feels the urgency. He has scheduled the debriefing for tomorrow. This afternoon, he will meet with his supervisor to review his audit findings and slides. Carlo welcomes his supervisor's input, because she has good insight on auditing activities and always seems to anticipate the issues that will be of greatest concern to study teams. He also wants to ask her opinion about a few of his audit observations; he is unsure whether to classify them as major or minor.

By the time he leaves for lunch, Carlo has finished the bullet points on his presentation slides. He hopes that he will be able to draft most of his audit report before the meeting with his supervisor in the afternoon. However, Nancy sees him in the cafeteria and wants to talk to him about another matter. She says that she has asked the CQA director for assistance with a systems audit of a vendor her study team is using, and he assigned Carlo as the auditor.

Over lunch, Nancy explains that she has contracted a company that provides radiology reader services. The external company uses board-certified radiologists, who are examining the X-rays of the patients in Nancy's study and making independent clinical assessments. The X-rays are blinded and, to further ensure accuracy, three radiologists evaluate each image. However, Nancy and the team have become increasingly concerned about the vendor's performance. The vendor has repeatedly missed its deadlines for submitting the radiologists' evaluations, and some X-rays appear to have been lost. Nancy is concerned because the vendor's poor performance could jeopardize the study. On behalf of the study team, she requested the CQA department to conduct a systems audit of the vendor's procedures and facilities.

Carlo's department calls this a "for-cause" audit; that is, the audit is scheduled because the requester is concerned or has evidence that regulatory standards and requirements may have been violated. The potential impact of suspected regulatory violations gives for-cause audits priority over other scheduled CQA activities.

Nancy gives Carlo several specific examples of the team's recent experiences with the vendor and promises to send him a more comprehensive list by email. They agree with the CQA director's estimate that two auditors should be sufficient to handle this audit. Carlo will concentrate on the procedures involved with shipping, evaluating, and storing the X-rays, and another CQA auditor, who is an information technology expert, will concentrate on the vendor's electronic and database systems. Nancy wants to schedule a special study team meeting this week to brief the two auditors. Carlo promises to contact the vendor and make arrangements for the audit visit.

When Carlo returns to his desk, he checks his email, which he neglected all morning, and sees the message from the CQA director alerting him about his new assignment. He also sees another message from the director to all of the CQA staff, requesting volunteers to assist with an ongoing regulatory inspection. One of the CQA managers has been hosting the inspectors during their audit at CanDo Pharma's facilities, but he has been overwhelmed with requests for documents and other arrangements. Carlo recognizes this as a rare opportunity to participate in a regulatory inspection and would like to help.

As the afternoon progresses, Carlo is still writing his audit report when his supervisor arrives for their meeting. They spend about an hour in her office reviewing his slides and discussing the strategy for his presentation. She reminds Carlo that he should be sensitive to the feelings of the study team members, who are working hard and trying to comply with all GCP requirements. They will react unfavorably to any criticism that they think is unfair. She makes several suggestions for changing the tone of the bullet points to make them more objective and lessen the possibility of offending the team.

Carlo also takes advantage of this opportunity to mention his new assignment on the for-cause audit and his desire to assist with the regulatory inspection. His supervisor is supportive of both activities, because they will strengthen his CQA credentials; however, she is concerned that he might not be able to meet all of the associated deadlines. Carlo must follow through with the audit he just completed and make sure that his final audit report is not delayed. She recommended him for the for-cause audit and is confident he will do a good job. The regulatory inspection is important, too, but she says other CQA workers have already offered to help. She thinks Carlo should concentrate on his two audit assignments and assures him that he will have other opportunities to participate in a regulatory inspection.

After their meeting, Carlo finishes the changes to his slides. Hopefully, the study team will not discover any misinterpretations in his audit findings. If the discussion goes smoothly tomorrow, Carlo can complete his draft audit report without major adjustments and quickly submit it to his supervisor for review—the first step in finalizing the report. He can then turn his attention to the for-cause audit.

HOW CARLO GOT HIS CQA AUDITOR JOB

Requirements

During the summers of his college years, Carlo worked as a technician at an environmental testing laboratory. He learned to follow a protocol, collect data, and keep an accurate laboratory notebook. When state officials inspected the laboratory for renewal of its commercial license, Carlo cooperated with the inspectors and realized the importance of thorough and accurate recordkeeping.

After graduating with a bachelor's degree in biology, Carlo found a job assisting a study coordinator at a local medical center. (See Chapter 3 for information about finding study coordinator jobs.) The coordinator depended on him to manage the site's study files, patient charts, and drug accountability records. In addition, the study coordinator and the visiting clinical research associates (CRAs) courteously explained GCPs and answered his questions about the clinical protocols.

Carlo quickly earned more responsibility, but he enjoyed maintaining the study binders and regulatory documents more than his other duties. Not surprisingly, before each scheduled CRA visit, the PI asked Carlo to work with the site staff to review all of the study files, update the binders, and file missing documents. Carlo's comprehensive understanding

---

BOX 7-2. *Requirements for a CQA Auditor Position*

- At a minimum, bachelor's degree in life science, nursing, or pharmacy (for systems auditors, a background in computers or engineering is helpful)
- Clinical experience: 1 to 3 years of work experience in life science or a medically related field (e.g., study coordinator or CRA)
- Proficient with GCPs, ICH, and other regulatory requirements
- Familiarity with study design, data management systems, and documentation practices
- Ability to focus on details and adhere to standards
- Excellent oral, written, and presentation skills
- Strong interpersonal skills
- Ability to influence and negotiate
- Use of good judgment and decision making
- Ability to work independently
- Strong organizational skills
- Ability to handle multiple competing priorities and manage time efficiently
- Excellent computer skills
- Ability to function as a member of a global team

of regulatory requirements and his diplomatic approach to correcting errors impressed his coworkers and the CRAs who monitored his site.

Through discussions with the CRAs, Carlo realized that CQA auditor positions matched his interests and skills. CQA auditors combine the investigative skills of a detective, the control and discipline of a commercial pilot, and the discretion of a diplomat. They must pay attention to details, sometimes digging deeply through disorganized documents; follow a myriad of regulatory rules that restrict their actions; and yet interact with people whom they audit in a friendly, positive, and helpful manner. He also liked the idea of traveling to clinical sites and working with clinical teams in many development programs.

Finding a CQA Auditor Position

You can find entry-level CQA auditor positions at large sponsor companies and at large CROs. Small companies typically hire a consultant or use their regulatory group to conduct audits. Large sponsor and CRO companies typically have fully developed quality assurance departments with experienced staff and established procedures. They offer an entry-level auditor the best opportunities for gaining high-quality training, experience, and career advancement.

Most companies post job openings on their websites, which can easily be searched by job title. The job titles for entry-level CQA auditor positions vary considerably between companies, but look for titles such as CQA auditor, quality assurance professional, quality assurance specialist, quality compliance specialist, or GCP auditor. (Large companies also use auditors to conduct audits of animal laboratories [Good Laboratory Practice (GLP)] and manufacturing facilities [Good Manufacturing Practice (GMP)], but those positions require different skills than clinical quality assurance auditors.)

Quality assurance departments are located in the research and development (R&D) division of sponsor organizations. In most companies, CQA is aligned with the clinical division; the laboratory and manufacturing divisions maintain separate quality assurance units. However, even in companies that combine clinical, laboratory, and manufacturing quality assurance in one department, a subgroup of CQA specialists focuses exclusively on clinical quality oversight.

The hiring managers at both sponsor companies and CROs look for the same skills and qualities in their job candidates: knowledge of GCP and other regulatory requirements, understanding clinical studies, interpersonal skills, and communication skills. Carlo's training and experience gave him the necessary qualifications for an entry-level CQA position, but there are other ways. Some job-seekers gain relevant recordkeeping and auditing skills at nonprofit organizations. Because CQA auditors must have a thorough understanding of clinical studies and GCPs, some auditors have worked as CRAs or in clinical data management before moving into a CQA position. (See Chapters 4 and 5 for information about CRAs and data management positions, respectively.)

Landing the Job

All of the hiring managers ranked Carlo as a highly desirable CQA candidate. Although he had never been an auditor, he was familiar with clinical studies from his work supporting the study coordinators. He understood GCPs and other regulatory requirements, and he knew the importance of accurate recordkeeping. Furthermore, his supervisors and the CRAs who monitored his site confirmed that he was extremely detail-oriented, could coordinate work on multiple tasks, and was an effective but pleasant negotiator.

BOX 7-3. *Tips for Getting a CQA Auditor Position*

- Consider working as a study coordinator
- Obtain accredited training in compliance and GCPs
- Consider managing patient charts in a doctor's office
- Consider an administrative position at a non-profit organization
- Consider working as an IRB administrator
- Join a professional society such as SQA, DIA, or ACRP
- Search CQA job postings on websites of drug, medical device, and CRO companies
- Preferentially look for entry-level jobs at large CROs and biopharmaceutical companies

The hiring managers were less impressed with the candidates who competed with Carlo for entry-level positions, even though some of them were CRAs with industry experience. CQA auditors often work unsupervised when auditing clinical sites and vendors. Hiring managers, therefore, rejected candidates whose previous supervisors said they were untrustworthy or could not manage their time wisely. Auditors must also have excellent interpersonal skills and encourage people to cooperate with the audit, even when deficiencies in quality are highlighted. Hiring managers, therefore, also rejected candidates who had a history of being disrespectful or overly assertive.

Working for a Sponsor Company versus a CRO

Although Carlo could have taken an entry-level CQA position at a large sponsor company, he decided to accept an offer from a large CRO for his first CQA job. Most CROs have excellent training programs and set specific skill standards for their employees. Also, at a CRO, Carlo knew that he would quickly gain auditing experience with many types of clinical studies and see the clinical operations at many different sponsor companies.

Because Carlo had not previously worked as an auditor, his first assignments were mainly administrative, along with on-the-job training to build his auditing skills. He also had to pass competency tests for GCPs, ICH guidelines, ISO standards, and the CRO's SOPs. By observing his coworkers, he learned how to interact with those who are being audited, how to use audit checklists and templates, and how to track the responses of open audit reports. He also read audit reports written by experienced auditors to become familiar with the style, format, and content of good reports.

Carlo continued his training by co-auditing under the supervision of a senior CQA auditor. Initially, he learned how to conduct internal systems audits, such as the CRO's clinical procedures, working alongside other internal auditors. As he gained experience, he began coauditing clinical studies at the investigator sites of some of the CRO's clients. For these audits, Carlo audited the sites for compliance with the sponsor's SOPs. Therefore, before he could conduct his audit, Carlo had to first understand that client's SOPs. Through these assignments, Carlo quickly showed his auditing skills and his flexibility in evaluating clinical site compliance with the SOPs established by different sponsors.

On the recommendation of his CQA trainers, Carlo assumed responsibility for independently planning, preparing, conducting, and reporting CQA audits. He supplemented his auditing skills and knowledge of regulatory requirements by taking online quality compliance courses. He also joined a professional society. (The Society of Quality Assurance [SQA], the Association of Clinical Research Professionals [ACRP], and the Drug Information Association [DIA] are three prominent societies for CQA specialists.) The advantages of membership include training classes, newsletters and other quality assurance-oriented publications, and an annual society meeting.

Carlo's career advancement in CQA depends heavily on his reputation as an auditor, and he can build his reputation most efficiently through his performance and networking. Good auditors earn the respect of their coworkers and the influential people whom they audit by demonstrating that they are thoroughly versed in all regulatory standards, apply those standards fairly and consistently, and help the clinical site personnel and study teams comply with those standards. Clinical teams would rather work with knowledgeable CQA auditors who conduct themselves in a cooperative manner as partners rather than as policemen and are more likely to recommend the advancement of auditors who help them.

Through his auditing assignments, Carlo meets many people, not only within his company but also well-known PIs, vendor personnel, and regulatory officials. If he impresses them with his auditing skills, regulatory knowledge, and professionalism, they can provide influential references and recommend him for promotion. Carlo's listing in his CQA society's membership directory also makes him visible to recruiters and hiring managers who are looking for experienced CQA auditors.

Although Carlo performed well and could have advanced his career at the CRO, he eventually decided to seek a position at a sponsor company. The salary and benefits are good at CROs, but the compensation packages at sponsor companies are often better. Also, Carlo wanted to gain experience with other types of audits (such as vendor audits), which are not typically handled by CROs.

Many sponsor companies prefer to hire experienced CQA auditors, because they want newly hired employees to begin auditing immediately. They draw heavily from CROs, which typically have qualified CQA auditors. Because Carlo was listed in the membership directory of his quality assurance society and several PIs spoke highly of Carlo's auditing skill, the

CanDo Pharma hiring manager easily identified him as a potential candidate.

By the time Nancy's study started, Carlo had conducted a number of internal and external audits at CanDo Pharma. When his supervisor selected Nancy's study for an audit, Carlo was fully prepared to conduct it.

MOVING FORWARD

Carlo's experience as a CQA auditor gives him many choices for career advancement, either at CanDo Pharma, another sponsor company, or a CRO. If he wants to stay in the quality field, he could continue gaining auditing experience and advance to more responsible positions in clinical quality assurance.

Titles for positions in the CQA auditor track differ between companies. They may be listed as Quality Assurance Auditor I/II/III, Quality Assurance Specialist I/II/III, or Auditor/Senior Auditor. Based on skills and performance, Carlo could earn promotions from one level to the next every 2–3 years. With each promotion, Carlo gains more responsibility and independence in conducting his audits. His audit assignments are larger, and sometimes he would lead a team of auditors for systems audits. Carlo also would have more opportunities to conduct global audits of clinical sites in multiple countries.

Carlo could also take additional training to learn specialized auditing skills. Regulatory authorities continue to set new and stricter requirements for using computers and electronic devices in clinical research. Conducting computer systems validations and audits of procedures such as electronic signature security and biostatistics methods requires special training. These specialized auditing skills are in great demand and include attractive compensation and benefits.

From the top of the CQA auditor career track, Carlo can move into a management position. Like managers in other clinical functions, Carlo would have supervisory responsibilities for a staff of CQA auditors, including their training, audit assignments, and performance reviews. In addition, CQA managers play an important role in managing clinical SOPs, quality assurance consulting, compliance training, and hosting regulatory inspections.

Regulatory authorities require sponsors to maintain up-to-date clinical SOPs. Each SOP is written by representatives who know and use the procedures they are describing. Data managers, for example, write the SOPs describing standard data management procedures. The CQA department serves as the clearing house for clinical SOPs, ensuring that the clinical authors include the correct information, cataloging and storing the SOPs in secure locations, and ensuring that the authors review and update the SOPs on time. The CQA managers therefore work closely with their clinical counterparts. Some companies use CQA experts in a separate standards unit that coordinates SOP management. (See Chapter 11 for more information about SOP standards management.)

Because they are experienced auditors, CQA managers know GCPs, regulatory requirements, and the sponsor's SOPs better than anyone else involved with clinical studies. Therefore, they often serve as consultants and respond to compliance questions from clinical teams and the company's executives. In addition, they proactively inform the clinical teams about newly published regulatory requirements and newly issued SOPs.

Regulatory agencies require all clinical team members to have adequate training for their jobs, and their employers must document those training records. In some companies, the CQA managers conduct compliance training on regulatory requirements, SOPs, and GCPs and maintain the training records for the clinical staff. In other companies, full-time training managers with the same background and knowledge handle these responsibilities. (See Chapter 11 for more information about clinical training jobs.)

In addition to the audits conducted by the CQA auditors, regulatory authorities send auditors to inspect clinical operations at sponsor companies and at the investigator sites. Sometimes, regulatory inspectors arrive unannounced. CQA managers host the regulatory inspectors at the sponsor's facilities and assist by retrieving documents, answering questions, and facilitating any other arrangements that the inspectors request during the inspection.

If Carlo is interested in advancing his career in quality management, he might pursue a position as the head of a quality assurance department. In addition to supervising the staff, quality assurance directors and vice presidents develop quality assurance strategies and may implement new quality assurance procedures. Some sponsor companies combine the various quality functions under a single director or vice president of quality or compliance. In this organizational structure, the quality function combines clinical quality assurance with the units responsible for laboratory and manufacturing quality. The quality assurance department may also pursue one or more corporate quality strategies such as total quality management (TQM), six sigma, ISO 9001, or ISO 13485 certification. To qualify for these senior-level positions, Carlo would need to broaden his auditing knowledge of GLP and GMP requirements, in addition to GCPs.

Finally, with his auditing experience, Carlo could advance his career by moving in a different direction, outside of CQA. He is well qualified to take a position in regulatory affairs (see Chapter 8), project management, standards, or training (see Chapter 11). If he is interested and is a talented writer, he can also consider a position in medical writing (see Chapter 10).

Whether Carlo chooses to continue working in the quality field or move to the regulatory affairs or project management career tracks, he can advance to positions that offer higher salaries, greater company visibility, more decision-making authority, budget responsibilities, and greater corporate benefits such as bonus and stock option packages. Rather than day-to-day operations, he would contribute to the strategic planning, oversight, and decisions regarding new product development. Experienced clinical trainers and medical writers also enjoy excellent compensation and benefits and can control their workload and assignments. However, their visibility and corporate influence stem from applying their skills rather than their strategic planning and decision-making authority.

CLINICAL QUALITY ASSURANCE RESOURCES

Compliance Standards and Training

International Conference on Harmonization (www.ich.org) sets international standards for clinical studies of investigational drugs.

International Organization of Standardization (www.iso.org) sets international quality standards recognized by governments for commercial products.

ISO 14155: General requirements for medical devices in clinical studies

ISO 14971: Risk management of medical devices

ISO 13485: Quality management of medical devices for regulatory purposes

Food and Drug Administration (www.fda.gov) sets safety regulations for foods, drugs, and other medical products in the United States.

European Medicines Agency (www.ema.europa.eu) evaluates and supervises safety regulations of drugs in the European Union.

European Commission/medical devices (http://ec.europa.eu/enterprise/sectors/medical-devices/index_en.htm) sets regulatory standards for medical devices in the European Union.

Compliance Online (www.complianceonline.com) offers online regulatory compliance training courses on pharmaceutical, medical device, ISO 9000, and clinical topics.

ClinfoSource (www.clinfosource.com) offers online courses on clinical quality assurance, clinical SOPs, and site audits for CME and CNE credit at reasonable cost.

Kriger Research Center International (www.krigerinternational.com) offers an online certificate program in quality assurance in multiple languages on the country-specific web pages under the "Courses" tab.

Good Clinical Practice Training and Certificate Programs (Including CME Credits)

Regulatory Affairs Professionals Society (www.raps.org) offers online GCP and clinical study management courses.

ClinfoSource (www.clinfosource.com) offers online GCP courses for CME and CNE credit at reasonable cost.

Kriger Research Center International (www.krigerinternational.com) offers online GCP courses and certification in multiple languages on the country-specific web pages under the "Courses" tab.

Association of Clinical Research Professionals (www.acrpnet.org) offers online GCP courses and certification for CME, CNE, and ACRP credit at reasonable cost.

American Society for Quality (www.asq.org) offers a certified quality auditor certification program.

GCP Training Online (www.gcptraining.org.uk) offers online GCP training and certification, emphasizing the European Clinical Trial Directive, at reasonable cost.

Clinical Quality Assurance Professional Organizations

Society of Quality Assurance (www.sqa.org) is an international professional organization promoting regulatory quality assurance of GLP, GCP, and GMP standards worldwide and providing a forum for information exchange and professional development of the quality assurance profession.

Drug Information Association (www.diahome.org) is a worldwide professional association that fosters innovation and exchange of health information. Members include those involved with discovery, development, regulation, and surveillance of biopharmaceutical products.

Association of Clinical Research Professionals (www.acrpnet.org) is an organization that provides educational and networking services for clinical professionals in the biopharmaceutical and medical device industries.

Profiles of Sponsor and CRO Companies (*See detailed list in Chapter 13*)

Pharmaceutical Research and Manufacturers of America (www.phrma.org) represents the leading pharmaceutical research and biotechnology companies in the United States.

Biotechnology Industry Organization (www.bio.org), the world's largest biotechnology organization, represents more than 1200 biotechnology companies.

Medical Design and Manufacturing (www.devicelink.com) is an online resource for the medical device industry.

Association of Clinical Research Organizations (www.acrohealth.org) represents the world's leading clinical research organizations.

Salary Surveys for Clinical Quality Assurance

American Society for Quality Salary Survey (www.asq.org) is published annually. Access the most recent survey by entering "salary survey" in the website's Search field.

American Association of Pharmaceutical Sciences Salary Survey (www.aaps.org) is published annually. Access the most recent survey by entering "salary survey" in the website's Search field.

Medical Device & Diagnostic Industry Salary Survey (www.devicelink.com) is published periodically. Access the most recent survey by entering "salary survey" in the website's Search field and selecting the Research and Development survey.

Salary.com (www.salary.com) publishes salary ranges for positions in a wide range of industries. Enter "quality assurance specialist" and, if appropriate, your targeted zip code in the Salary Wizard.

8

Entering Regulatory Affairs

A REGULATORY AFFAIRS SPECIALIST is the person responsible for managing the company's activities for regulatory approval of its products, compiling regulatory documents, communicating with regulatory authorities, and maintaining regulatory materials on marketed products. Although many regulatory agencies have cooperated and harmonized their drug and medical device requirements into international standards, many variations still exist, and all agencies periodically update, revise, and strengthen their country-specific requirements. Keeping up with the current versions of all those requirements is a full-time job, and the clinical staff relies on regulatory affairs specialists to keep everyone informed.

All biopharmaceutical and medical device companies have at least one regulatory affairs specialist to oversee the company's regulatory activities and serve as the official who communicates with the regulatory authorities. Typically, large sponsor companies use a sizable staff of regulatory affairs experts. Some regulatory affairs specialists, like Amy at CanDo Pharma, work directly with product strategy teams and clinical study teams as an advisor on regulatory issues. Other specialists work in a regulatory operations group and are responsible for generating, submitting, and archiving official regulatory documents.

Some contract research organizations (CROs) employ regulatory affairs consultants who offer their services to small sponsor companies that cannot afford a full-time regulatory staff. More frequently, sponsor companies will hire a CRO with regulatory expertise to compile and maintain their regulatory documents, but the sponsor's regulatory staff retains control of the regulatory strategy and communications with regulatory authorities.

BOX 8-1. *Regulatory Affairs Specialist Responsibilities*

- Provide regulatory affairs expertise on project and program teams
- Advise clinical study teams on regulatory issues
- Compile essential documents for regulatory submission
- Submit the protocol, final study report, and other required documents to regulatory agencies
- Communicate serious adverse events to regulatory agencies
- Compile and submit regulatory progress reports
- Communicate with regulatory authorities regarding meeting arrangements, submitted documents, and responses to regulatory requests
- Track all discussions and file all correspondence between the sponsor and regulatory agencies
- Coordinate preparation, submission, and maintenance of regulatory documents for review and approval, such as INDs, IDEs, MAAs, and PMAs
- Ensure that the elements, quality, accuracy, and format of regulatory submissions comply with laws, regulations, and corporate standards
- Perform regulatory review of product labeling, claims, and advertising
- Cooperate with regulatory authorities during regulatory inspections
- Monitor, analyze, and advise sponsor on regulatory trends and emerging requirements

THE ROLE OF THE REGULATORY AFFAIRS SPECIALIST

Amy's interactions with regulatory agencies and the groups at CanDo Pharma are illustrated in Figure 8-1. Amy does not have supervisory responsibilities, but the regulatory affairs department provides a few administrative assistants to handle clerical and administrative tasks. Amy can also rely on the services of the regulatory department's operations group to assist her with document production, submission, and archiving. Her purchasing ability is limited by the expense authority level associated with her job title. However, the regulatory department's annual budget incorporates the costs of her job-related expenses, including travel.

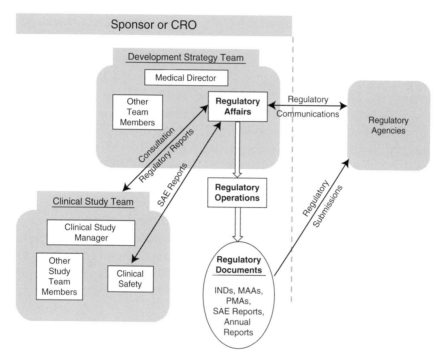

Figure 8-1. Regulatory affairs specialist interactions with regulatory agencies and sponsor company.

Investigational New Drug/Clinical Trial Authorization Preparation

Amy has been a member of Dr. Abernathy's strategy team from the beginning of the development program on CanDo Pharma's experimental drug. Sponsors cannot initiate clinical studies in any country until they comply with the human experimentation requirements of that country's regulatory agency. Those requirements include data from animal studies. Amy advised the strategy team and answered questions about the regulatory requirements for conducting the preclinical laboratory studies using the CanDo Pharma drug. The data from studies in animals and other laboratory assays showed the drug's safety and potential medicinal value. Similar preclinical safety testing is required for Class III medical devices (such as nicotine patches, artificial hips, and pacemakers).

To prepare the application for authorization to start clinical testing, Amy worked with the strategy team to compile the accumulated laboratory results and information about the drug's chemical characteristics. They also used that data to prepare the first version of the investigator's brochure (IB). (See Chapter 3 for more information about IBs.) In addition, the team prepared the protocol for its first proposed clinical study and drafted the sample informed consent form.

Amy and her regulatory coworkers assembled the laboratory data, clinical protocol, and other documents into the application format required by the regulatory authorities. In the United States, she submitted an Investigational New Drug (IND) application for review by the Food and Drug Administration (FDA). (For Class III medical devices, the FDA requires an Investigational Device Exemption [IDE] application.) Member countries in the European Union (EU) require a Clinical Trial Authorization (CTA) application, which has documentation requirements similar to an IND, but each country specifies slightly different submission procedures. (See Chapter 2 for details of regulatory documents and procedures.)

Ongoing Regulatory Communications

Amy is the person responsible for interactions with the regulatory authorities on the team's behalf, and the regulatory authorities direct their comments and requests for information to her. The communications are ongoing and actually begin prior to submitting the IND and CTA applications to start clinical studies. The types of interactions include teleconferences, meetings, and informal discussions concerning CanDo Pharma's regulatory applications, amendments, and product development plans. Amy archives summaries of all her interactions with the regulatory authorities and keeps copies of their inquiries, requests, and comments in her departmental files. Amy also complies with regulatory requirements for archiving copies of all regulatory applications, amendments, reports, and correspondence.

After CanDo Pharma received authorization to began clinical testing, Dr. Abernathy's strategy team planned a series of further laboratory and clinical studies to complete the development program on the CanDo Pharma drug. Mindful of the regulatory requirements, Amy dutifully informed the regulatory authorities of each significant new activity in CanDo Pharma's development program. She submitted the protocols for each new clinical study, IB updates, and annual summaries of all accumulated data. All of the submissions to the FDA referenced the original IND application and became updates to the FDA's file on CanDo Pharma's experimental drug. For each new clinical study that CanDo Pharma conducted in European countries, Amy submitted a new CTA and obtained a unique EudraCT number, which registered the study in the EU clinical study database. (See Chapter 2 for more information about IND and CTA application procedures.)

Planning a New Clinical Study

By the time that Nancy, the clinical study manager, David, the clinical research associate (CRA), and the other team members began planning their clinical study, Amy was already familiar with the CanDo Pharma drug from the results of previous preclinical and clinical studies. Some regulatory authorities had raised questions about the earlier clinical data, and Amy shared this information with Nancy's clinical study team. For example, some patients developed a skin rash during treatment, but Dr. Abernathy was uncertain whether drug treatment had caused it. Consequently, the clinical protocol for Nancy's study includes procedures to closely monitor all patients who develop skin rashes.

Because Nancy's study uses investigator sites in the United States, Australia, and Germany, Amy submits the final protocol to the regulatory agencies of all three countries. Regulatory procedures differ between countries. Some agencies such as the FDA in the United States review each new protocol. Other agencies add the new protocol to their existing file on the CanDo Pharma drug without review.

If the reviewer at a regulatory agency has concerns or questions about the protocol, Amy works with the team to address the problems. They may need to explain the study procedures in greater detail or perhaps change the study design. Amy responds to the agency on behalf of the team and, if indicated, submits an amended protocol. Because principal investigators (PIs) at all sites must conduct the study in exactly the same way, Amy submits the amended protocol to all three regulatory agencies, and the team instructs all of the PIs to follow the new version.

Regulatory agencies require sponsors to notify the general public of their ongoing and completed clinical studies. For example, the FDA requires sponsors to post study notifications on the www.Clinical Trials.gov website. In Europe, the Medical Device Directive requires similar notifications for clinical studies of medical devices. These websites permit patients, healthcare professionals, and others to view information about clinical studies such as the study objectives, the experimental product being tested, eligibility criteria for patients, the investigator sites, and the sponsor.

Reporting Serious Adverse Events

Once Nancy's study starts, Amy's involvement is limited, because clinical teams do not need ongoing regulatory guidance. However, she readily responds to questions from the team about regulatory issues or procedures, proactively informs them when regulatory agencies issue new requirements, and notifies the regulatory agencies when serious adverse events (SAEs) occur.

SAEs require the team's and Amy's immediate attention. Regulatory agencies specify precise procedures and a rapid timetable for reporting an SAE. Amy works closely with Dr. Abernathy and Jasmine, the clinical safety specialist, to collect details of the SAE, and Amy submits the SAE report to the regulatory agencies, according to each country's procedures. (See Chapter 9 for more information about handling SAEs.)

When they review SAE reports, the regulatory authorities sometimes request more information or raise concerns. In these cases, Amy coordinates the communications between the authorities and Nancy's team. When the regulatory issues are resolved, she archives all discussions, correspondence, and actions in the regulatory department files for future reference.

Annual Reports and Clinical Study Reports

Each year, Amy works with her colleagues in the regulatory operations group and Dr. Abernathy's strategy team to compile a safety report on the CanDo Pharma drug. This annual report includes a summary of the status of Nancy's ongoing clinical study and all other clinical studies of the drug, an updated IB, tables summarizing the observed adverse events (AEs) and SAEs, major findings of new animal studies, and manufacturing procedure updates. Although many people contribute information to the annual safety report, Amy coordinates their work and advises the regulatory operations group as they compile the final document. She submits the annual safety report to the regulatory agencies, each of which has different requirements for format and content.

After Nancy's study is completed, Mike, the medical writer, prepares the clinical study report with input from the team. (See Chapter 10 for more information about preparing clinical study reports.) Although Amy is not a coauthor, she reviews and approves the final report. Her signature indicates that the report meets the requirements for format and content of each regulatory agency. After all the CanDo Pharma representatives sign the final study report, Amy submits it to the regulatory agencies. Amy also ensures that the team complies with regulatory agency requirements for posting a summary of the study results on a public access website such as www.clinicalstudyresults.org.

Regulatory Inspections

Occasionally, regulatory agencies send auditors to inspect clinical operations at sponsor companies and at investigator sites. Clinical quality assurance (CQA) managers usually host the visiting inspectors and facilitate the inspection by retrieving documents and arranging interview schedules. (See Chapter 7 for details of CQA activities.) If the inspection focuses on the development program for the CanDo Pharma drug, Amy assists Dr. Abernathy's strategy team by providing regulatory guidance. She also retrieves documents from her files to answer questions about previous interactions between CanDo Pharma and the agency.

Violations cited by regulatory inspectors can have serious consequences for the company, its management, and the PIs who participate in its clinical studies. The authorities can impose fines, close manufacturing facilities, and file criminal charges. After an inspection, Amy works diligently with the team to address and respond to the inspection findings. She may not be directly involved with the decisions and corrective actions taken by CanDo Pharma employees and executives, but all discussions with the authorities during the inspection resolution process flow through Amy's office. She also archives summaries of the corrective actions and the discussions with regulatory authorities in her departmental files.

ONE OF AMY'S DAYS

Over her morning coffee, Amy surfs through a series of regulatory websites looking for notices of new and revised regulations. She also receives automatic e-mail messages from several organizations that monitor regulatory activities and send news alerts. The research and development (R&D) teams at CanDo Pharma rely on Amy and her regulatory colleagues to inform them about changes in the regulatory requirements that apply to their studies.

Because regulatory agencies follow a slow and methodical process when implementing new regulations, Amy knows well in advance about pending changes. However, she closely monitors the authorities' progress and, most importantly, the effective date of new rules. She takes the initiative to alert the CanDo Pharma employees who are affected by these regulations and makes sure that they have taken the appropriate steps when the rule goes into effect.

This morning, she is monitoring a clinical regulation that is being revised to incorporate additional requirements for reporting safety data. The text of the revised rule is in the final stage of government review, and the regulatory authorities aim to implement the new requirements at the end of the year. Amy has been keeping her teams informed about the regulators' progress, and the CanDo Pharma clinical department has been revising several of its SOPs to reflect the new requirements. The clinical teams would rather comply with the new regulations for clinical studies that are starting now, rather than make adjustments later in the middle of ongoing studies.

After she finishes reading the new notices, Amy leaves her office to attend a meeting with Dr. Abernathy's strategy team. Dr. Abernathy has approved the clinical protocol for a new clinical study in his development program, and the team discusses plans for starting the study. Amy's responsibility is to submit the final protocol to the regulatory agencies in all of the countries where the study will have investigator sites. To assist her, Amy will coordinate the submissions with her regulatory colleagues in CanDo Pharma's French and Australian offices. The French regulatory specialist will handle the submissions to each of the European countries involved with the study, the Australian specialist will submit the protocol to the Australian and New Zealand authorities, and Amy will submit it to the agencies in North America.

Amy tells the strategy team that she has alerted her colleagues about the protocol, and they are ready to submit it, along with the required supporting documents, to each country. She has also worked out a timetable with her colleagues, showing the estimated time for the authorities to complete their respective reviews. The timeframes vary, but she expects that it will take up to 6 months to receive authorization from the regulators in some countries.

When she returns to her desk, Amy sees an important message in her e-mail inbox: A copy of the inspection report that regulatory auditors sent to one of CanDo Pharma's PIs. For several years, the PI had participated in CanDo Pharma clinical studies and had always enrolled large numbers of patients. However, in the most recent study, the clinical research associate (CRA) monitoring his site probed deeper into the clinical site's medical records and discovered some inconsistencies. For example, the medical charts for many patients did not list a diagnosis for the disease under study until the time the patient enrolled in the study. For other patients, the documents showing that the patients met the study's inclusion and exclusion criteria appeared to be postdated. When the CRA asked for clarification on these issues, the PI became increasingly uncooperative. CanDo Pharma's executives eventually, and reluctantly, terminated the PI's contract. In addition, out of a concern for the welfare of patients who were participating on studies by other sponsors at this clinical site, CanDo Pharma informed the regulatory authorities about its experiences.

The regulatory authorities had quickly reacted and deployed auditors to inspect the site, and Amy anxiously waited for the inspection report. Indeed, the inspectors detailed a number of serious violations of applicable laws and regulations, for which the PI was unable to give adequate explanations. Amy forwarded copies of the inspection report to Dr. Abernathy, his medical colleagues, and other members of CanDo Pharma's senior management. She knew they would take immediate steps to protect the integrity of CanDo Pharma's clinical data. That meant systematically going through each of the studies on which this PI had participated and segregating the data associated with patients enrolled at his site.

She hoped they would determine that some of the data were still valid. However, in many cases they would probably conclude that the patients were ineligible and the corresponding data would have to be removed from the final data analyses. Because of the large number of patients involved, the clinical data remaining in some studies might not be sufficient to draw statistically valid conclusions, and those studies would need to be repeated.

These were serious concerns with widespread repercussions. Amy spent the rest of the day responding to questions from the affected clinical study teams and her regulatory colleagues regarding the steps they should take, or were required to take, in response to the inspection findings. Her manager suggested that she, as the most knowledgeable regulatory specialist on the studies in which this PI participated, should take the lead in setting up a series of meetings to explain the inspection report findings, clarify CanDo Pharma's obligations, and discuss the actions that the various clinical teams should take to preserve the integrity of their data.

Because everyone was anxious about the impact of this PI's misconduct and rumors were already flying around the company, Amy scheduled meetings with all of the affected teams, starting tomorrow morning. She knew she would be asked about CanDo Pharma's regulatory obligations, liability, and options. So, she worked late rereading the relevant regulatory publications, discussing regulatory strategies with the head of the regulatory affairs department, and preparing summary slides of the inspection report findings.

HOW AMY GOT HER REGULATORY AFFAIRS JOB

Requirements

After Amy received her bachelor's degree in general science, she took a clerical job at a nearby medical center. Her duties included retrieving and filing patient charts, processing laboratory test results, and scheduling medical procedures. Through these activities, Amy learned medical terminology and became acquainted with clinical laboratory procedures.

Because of staff shortages, Amy volunteered to assist several study coordinators with their clinical study documents. She liked working with the coordinators, and they appreciated her ability to organize and update their study files. After Amy took an online training course on Good Clinical Practices (GCPs), she asked the coordinators for additional assignments. They were grateful for her help, especially when they prepared for monitoring visits by CRAs.

Through discussions with the visiting CRAs, Amy first learned about regulatory affairs. She enjoyed learning about regulatory requirements such as GCPs and International Conference on Harmonization (ICH) guidelines, and she appreciated the importance of following rules,

BOX 8-2. *Requirements for a Regulatory Affairs Position*

- At a minimum, bachelor's degree in science, pharmacy, or nursing (for medical device jobs, a background in engineering or manufacturing is an advantage)
- Clinical experience: 1 to 3 years of work experience in life science or a medically related field (e.g., clinical auditing)
- Knowledge of the global regulatory affairs environment
- Fully versed in GCPs, ICH, and other regulatory requirements related to quality and compliance
- Ability to understand and interpret government regulations
- Experience with document management systems (e.g., Documentum, CoreDossier)
- Strong computer skills (MS Word, Excel, Project, PowerPoint)
- Excellent interpersonal, verbal, and written communication skills
- Ability to organize, prioritize, and implement multiple assignments
- Ability to meet deadlines
- Flexible and adaptable to changing priorities
- Attention to detail
- Problem solving ability
- Ability to work independently

especially when the experimental studies involved patients. She also liked the idea of interacting with regulatory authorities. Because the structure and attention to detail associated with regulatory work appealed to her, Amy realized that regulatory affairs was a good career move for her.

Finding a Regulatory Affairs Position

All biopharmaceutical and medical device companies and many CROs have a regulatory affairs function. Most of these companies post job openings on their websites, which can easily be searched by job title. The job titles for entry-level regulatory affairs positions vary considerably between companies, but look for titles such as regulatory affairs associate, regulatory affairs specialist, and regulatory affairs professional.

Large sponsor companies and many large CROs have fully developed regulatory affairs departments. The experienced staff, established procedures, and validated document management systems offer an entry-level worker the best opportunities for gaining high-quality training, experience, and career advancement.

The regulatory affairs departments at large companies are usually subdivided into several functional units. One or more units work directly with the development teams to provide regulatory advice and interact with regulatory authorities. The other unit, regulatory operations, compiles and publishes the company's regulatory documents. Because of the strict regulatory requirements for reporting SAEs, some companies also include the clinical safety function (see Chapter 9) within the regulatory affairs department.

At small sponsor companies, regulatory affairs workers are less specialized and have a wide range of duties. Small biopharmaceutical companies typically hire only experienced regulatory affairs managers, who do not require extensive training and supervision. Small medical device companies often rely on regulatory consultants, rather than maintaining an internal regulatory affairs unit.

The work of regulatory affairs specialists in the medical products industry is unique. Hiring managers at both CROs and sponsor companies, therefore, do not expect entry-level workers to have previous regulatory experience with medical products; instead, they look for candidates who have applicable skills and qualities, such as knowledge of GCP and ICH guidelines, familiarity with drug (or medical device) development, and the ability to work simultaneously on overall objectives and fine details. Entry-level workers usually start in the regulatory operations group, unless they already have a strong scientific background and a good understanding of the drug (or medical device) development process.

Amy's experience as a medical records clerk and assisting the study coordinators gave her the necessary qualifications for an entry-level regulatory affairs position, but there are other ways. Some universities offer certificates or degree programs in regulatory sciences. Although these university programs will not give you regulatory experience, they do provide a firm foundation on regulatory principles. Study coordinators gain relevant experience by reading clinical protocols, following GCPs, submitting SAE reports, and maintaining regulatory documents in their study files. (See Chapter 3 for details on landing a study coordinator job.) Many regulatory affairs specialists at medical device companies have a background in engineering or manufacturing.

Another approach is to serve as the administrative assistant for a local institutional review board (IRB). The IRB administrator is a very responsible person, handling all of the clinical study documents on behalf of the IRB's chairman, and this is a good way to learn about GCPs, study protocols, informed consent issues, and other regulatory procedures.

For regulatory affairs positions in the product support group, entry-level candidates must have a strong scientific background, project management skills, an understanding of regulatory requirements, and knowledge of the drug (or medical device) development process. Therefore, product-oriented regulatory affairs positions at sponsor companies are usually filled by internal candidates, because they already know the company, its products, and product development procedures. Effective strategies for landing an entry-level, product-oriented regulatory job include

---

BOX 8-3. *Tips for Getting a Regulatory Affairs Position*

- Take online courses on GCP and the development process for new drugs or medical devices
- Learn medical terminology
- Review regulatory requirements posted on the ICH and FDA websites
- Consider learning document management system software such as Documentum
- Consider jobs such as study coordinator, IRB administrator, or essential document assistant, which include handling clinical study documents
- Consider jobs such as medical records clerk or medical office assistant that include handling patient charts, clinical laboratory test results, and other medical records
- Consider a university certificate or degree in regulatory sciences
- Consider working at a regulatory agency
- Join a professional society such as RAPS
- Seek certification as a regulatory affairs professional such as RAC
- Search regulatory affairs job postings on websites of drug, medical device, and CRO companies
- Preferentially look for entry-level jobs at large CROs and biopharmaceutical companies

experience as a CRA (see Chapter 4) or a project planner in the project management unit (see Chapter 11).

Landing the Job

All of the hiring managers ranked Amy as a highly desirable regulatory candidate. She was familiar with medical recordkeeping, medical terminology, and routine laboratory tests. Her coworkers confirmed that she was meticulous in processing laboratory results, filing patient charts, and scheduling patient visits. The hiring managers were impressed that Amy took the initiative to learn about GCPs and other relevant regulations. In addition, her work assisting the study coordinators gave the hiring managers confidence that she understood clinical studies. The study coordinators and visiting CRAs enjoyed working with her and were impressed with her ability to prioritize multiple tasks and meet their deadlines.

The hiring managers were less impressed with the candidates who competed with Amy for entry-level positions, even though some were study coordinators. Regulatory affairs specialists often represent their company in interactions with regulatory authorities, and they regularly participate in internal team and management discussions. The hiring managers therefore rejected candidates who did not show good written and oral communication skills and those who had a history of poor teamwork and interpersonal skills. Also, because regulatory affairs specialists must comply with rigorous deadlines imposed by regulatory agencies, the hiring managers also rejected candidates who did not reply to requests or submit the supporting documents for their job applications in a timely manner.

Working for a Sponsor Company versus a CRO

With her background, Amy's best opportunities for an entry-level regulatory affairs position were at large CROs. CROs offer considerable exposure to a range of regulatory experiences, product development activities, and therapeutic areas.

Under the supervision of her CRO manager, Amy successfully completed the CRO's required regulatory training modules, including additional GCP training, ICH and country-specific regulatory requirements, and the CRO's standard operating procedures (SOPs) for conducting regulatory affairs activities. She also earned her certification on the document management system that the CRO used to compile marketing authorization applications (MAAs) and other documents for regulatory submission. Increasingly, sponsors submit their regulatory documents electronically, and regulatory agencies specify the standards for these electronic common technical documents (eCTDs). Two document management systems commonly used by sponsor companies are Documentum and CoreDossier, because they comply with the eCTD requirements.

Initially, Amy worked under the supervision of the CRO's regulatory affairs manager to handle the essential documents of one of their clients' study teams. Essential documents are documents that must be generated and maintained by the investigator, sponsor, or monitor in support of a clinical study. These documents, such as the clinical protocol, IB, informed consent form (ICF), IRB approval letter, drug shipping records, and SAE reports, verify that the team is complying with GCPs and all other applicable regulatory requirements. Amy quickly became an expert in knowing which documents to collect, when to collect them, where to store them, and how and when to submit them to the regulatory agencies.

As she gained experience, she took more responsibility for reviewing the essential documents to make sure they were completed correctly and advising the PIs and study teams accordingly. In conjunction with her work, she also became familiar with the procedures and activities of site monitors, IRB practices, and the interdisciplinary activities involved with new product development. Because she was qualified to use the regulatory department's document management system, she assisted the operations group with compiling several documents for regulatory submission.

During this time, Amy joined a professional society for regulatory affairs specialists. (Regulatory Affairs Professional Society [RAPS] is a prominent society for regulatory affairs professionals.) The advantages of membership include training classes, an official certification program, newsletters and other regulatory affairs-oriented publications, and an annual society meeting.

The most effective ways for Amy to advance her career as a regulatory affairs professional are to build her reputation as a regulatory expert, earn regulatory affairs certification (RAC), and maintain her listing in her regulatory society directory. Regulatory affairs professionals earn the respect and confidence of the influential people in their company by knowing the nuances of all applicable regulations and developing a productive rapport with regulatory authorities. Many highly valued regulatory affairs specialists acquire that type of experience by working as an employee at a regulatory agency. In addition, Amy should take advantage of every opportunity to strengthen her scientific and medical knowledge, so that she understands the therapeutic and adverse actions of products under development.

Certification is an additional qualification that Amy can add to her résumé. It gives her an advantage over other candidates if she decides to apply for regulatory affairs positions at other companies. RAPS awards RAC certification to members who have shown their knowledge of US, EU, or Canadian regulations through training, experience, and an examination. Amy's listing in her regulatory society membership directory also makes her visible to recruiters and hiring managers who are looking for knowledgeable regulatory affairs professionals.

Amy performed well and could have advanced her career to more responsible regulatory affairs positions at the CRO. However, her opportunities for submitting documents and interacting with regulatory authorities were limited. Sponsor companies prefer to interact directly with regulatory agencies, and most prefer to submit their regulatory documents directly to the agencies, even when a CRO prepared them on the sponsor's behalf. Although Amy liked the CRO, she saw greater opportunities for regulatory career advancement at a sponsor company.

Because Amy was listed in the membership directory of her regulatory affairs society, the CanDo Pharma hiring manager easily identified her as a potential candidate, and her regulatory affairs experience at the CRO gave her the necessary prerequisites. At CanDo Pharma, Amy continued accumulating regulatory expertise and quickly learned about CanDo Pharma's experimental products. In addition to helping the development teams compile regulatory documents for submission, she arranged meetings with regulatory authorities and participated in teleconferences with regulatory reviewers.

When Dr. Abernathy's strategy team began planning the development of the CanDo Pharma drug, Amy became the team's regulatory representative. By the time Nancy's clinical study started, Amy had advised Dr. Abernathy on a number of regulatory issues and submitted all of the preceding regulatory documents on the strategy team's behalf. She was familiar with the CanDo Pharma drug and the responses of the regulatory authorities regarding the earlier clinical studies. She was fully prepared to advise Nancy's study team on the regulatory aspects of their work.

MOVING FORWARD

Amy decided to continue her career in regulatory affairs. From an entry-level position, regulatory affairs specialists can be promoted to several additional levels of regulatory responsibility. Companies differ in the titles for these levels. Some use a numerical scale (e.g., regulatory specialist I, II, and III) and others use gradated titles (e.g., regulatory affairs associate and senior associate). Depending on performance and the company's structure, Amy can expect a promotion every few years to the next level of responsibility. After sufficient experience, Amy could qualify for promotion to a regulatory affairs manager. Some regulatory affairs managers primarily manage a staff of regulatory specialists, usually in the regulatory operations group. Other managers support the development strategy teams without supervisory responsibilities.

Amy's training on electronic document management systems at the CRO and her other regulatory experience give her the qualifications to manage a regulatory operations group, which is responsible for compiling regulatory documents such as Investigational INDs, IDEs, and MAAs. (See Chapter 2 for details of regulatory document requirements.) Most sponsors and regulatory agencies prefer electronic regulatory documents using the eCTD format rather than paper copies. Therefore, managers of the regulatory operations group must have strong computer skills and a thorough understanding of a validated document management system, such as Documentum.

In large regulatory operations groups, some workers specialize in certain aspects of operations activities, such as registration coordinators, archive specialists, or electronic submissions specialists. As the manager of this group, Amy would be responsible for training the regulatory affairs specialists, assigning their document preparation tasks, and evaluating their performance.

However, Amy preferred to continue working with the development teams as a senior regulatory advisor. Unlike managers of regulatory operations groups, regulatory managers who work closely with clinical teams have no, or only a few, regulatory specialists reporting to them. Amy's regulatory experience makes her a valued consultant to the teams and liaison with the regulatory agencies. Development teams especially welcome the knowledge of regulatory managers when they prepare MAAs for regulatory review and approval.

As a regulatory manager, Amy would plan and coordinate the work of Dr. Abernathy's strategy team to prepare the MAA (called a New Drug Application [NDA] in the United States) for the CanDo Pharma drug. In addition to the individual study reports, the Common Technical Document (CTD) format requires an overall summary of the drug's therapeutic benefit (i.e., clinical summary of efficacy), an overall summary of the drug's side effects (i.e., clinical summary of safety), reports and summaries of the laboratory studies in animals, and a variety of documents describing drug manufacturing.

The MAA for approval of a Class III medical device (called a Premarket Approval [PMA] in the United States) requires a similar compilation of preclinical and clinical data. In addition, the sponsor must provide detailed information describing how the device is manufactured.

In conjunction with the regulatory operations group and her colleagues in CanDo Pharma's regional offices, Amy would submit the completed MAA to the regulatory agencies in each country where CanDo Pharma seeks market approval. She would keep a complete copy of the MAA in the regulatory department files and track all discussions between the agencies and CanDo Pharma during the review and approval process. Typical interactions between regulatory agencies and sponsors prior to and during MAA review include meetings, inspections, and product labeling discussions.

Meetings between sponsors and regulatory authorities are held to discuss specific issues and may either take place at the agency or via teleconference. Meetings at key times during the clinical development program allow the sponsor to discuss its development plan and study designs, gauge the regulatory authorities' views, and make adjustments to keep the development program properly focused. Meetings prior to MAA submission (called the pre-NDA meeting in the United States) allow the sponsor to ensure that it will meet the regulatory requirements for data format in the submission. Meetings during the MAA review process allow the sponsor to explain or interpret data when the regulatory reviewers express concerns or have questions about the content of the application. For all meetings with the regulatory authorities, Amy and the agency contact person would coordinate meeting arrangements such as the meeting date, agenda, premeeting documents, and the meeting minutes.

As part of the MAA review process, some agencies conduct an inspection of the sponsor's facilities, contractor's facilities, or investigator sites. For example, the FDA routinely inspects the facilities where the sponsor plans to manufacture commercial batches of a new drug. For these "preapproval" inspections, which are usually scheduled by the regulatory agency, Amy would coordinate arrangements between her contact person at the agency and the appropriate people at CanDo Pharma. During the inspection, which is usually hosted by CanDo Pharma's quality assurance managers, Amy would assist the CanDo Pharma team by providing documents from the regulatory files, answering questions, and verifying previous team activities. Following the inspection, Amy would work with the team to respond to inspection findings and, if indicated, document corrective actions.

As a regulatory manager, Amy also would mediate agreements between CanDo Pharma and the regulatory agency regarding the product's label. The label, which includes the text and design of the product package and additional product description leaflets placed inside the package, must accurately summarize the features of the product, how to administer it, special precautions and warnings, and instructions for its use. Amy would work with the team to coordinate the response to the agency's comments and counterproposals until CanDo Pharma and the agency agree on the final label design and text.

At a senior level, Amy might eventually qualify to become the head of a regulatory affairs department. Her promotion to this level would depend on her expertise as a regulatory affairs professional, her ability to think strategically, and her leadership skills. Sponsors depend on the regulatory affairs department head to advise the company's executives on regulatory risks and strategies for responding to regulatory challenges. Regulatory affairs leaders also set the tone for the company's conduct and compliance with regulatory requirements.

If Amy prefers to advance her career in another direction, her regulatory affairs experience gives her the qualifications to pursue a variety of career opportunities. She is well qualified to take a position in project management, compliance training, or quality standards management (see Chapter 11 for details of these career options). If she is interested and is a talented writer, she can consider a position in medical writing (see Chapter 10).

Whether Amy chooses to continue working in regulatory affairs or move to the project management or quality standards career tracks, she can advance to positions that offer higher salaries, greater company visibility, more decision-making authority, budget responsibilities, and greater corporate benefits, such as bonus and stock option packages. Like her work as a regulatory affairs specialist, she would contribute to the strategic planning, oversight, and decisions regarding new product development.

REGULATORY AFFAIRS RESOURCES

Compliance Agencies and Standards

International Conference on Harmonization (www.ich.org) sets international standards for clinical studies of investigational drugs.

International Organization of Standardization (www.iso.org) sets international quality standards recognized by governments for commercial products.

ISO 14155: General requirements for medical devices in clinical studies

ISO 14971: Risk management of medical devices

ISO 13485: Quality management of medical devices for regulatory purposes

Food and Drug Administration (www.fda.gov) sets safety regulations for foods, drugs, and other medical products in the United States.

European Medicines Agency (www.ema.europa.eu) evaluates and supervises safety regulations of drugs in the European Union.

European Commission/medical devices (http://ec.europa.eu/enterprise/sectors/medical-devices/index_en.htm) sets regulatory standards for medical devices in the European Union.

Good Clinical Practices Training and Certificate Programs (Including CME Credits)

Regulatory Affairs Professionals Society (www.raps.org) offers online GCP and clinical study management courses.

ClinfoSource (www.clinfosource.com) offers online GCP and product development courses for CME and CNE credit at reasonable cost.

Kriger Research Center International (www.krigerinternational.com) offers online GCP courses and certification in multiple languages on the country-specific web pages under the "Courses" tab.

Association of Clinical Research Professionals (www.acrpnet.org) offers online GCP courses and certification for CME, CNE, and ACRP credit at reasonable cost.

GCP Training Online (www.gcptraining.org.uk) offers online GCP training and certification, emphasizing the European Clinical Trial Directive, at reasonable cost.

Medical Dictionaries and Terminology

ClinfoSource (www.clinfosource.com) offers an online medical terminology course for CME and CNE credit at reasonable cost.

Kriger Research Center International (www.krigerinternational.com) offers an online course in medical terminology in multiple languages on the country-specific web pages under the "Courses" tab.

Free-ed.net (www.free-ed.net/free-ed/healthcare/medterm-v02.asp) offers an online course in medical terminology and CEU credit at reasonable cost.

Des Moines University (www.dmu.edu/medterms) offers a public access, online course in medical terminology.

Regulatory Affairs Courses and Certificate Programs (Including CME)

Regulatory Affairs Professional Society (www.raps.org) offers online regulatory courses and Regulatory Affairs Certification (RAC) programs for US, EU, and Canadian regulations.

Compliance Online (www.complianceonline.com) offers online regulatory compliance training courses on pharmaceutical, medical device, ISO 9000, and clinical topics.

ClinfoSource (www.clinfosource.com) offers online courses on IND and NDA submissions, the European Clinical Trial Directive, and the Health Canada regulatory framework for CME and CNE credit at reasonable cost.

Medical Design and Manufacturing (www.devicelink.com) offers a wide variety of archived free webcasts on topics of importance to professionals in the medical device industry.

Document Management Systems (Including Training and Certification)

Documentum (www.emc.com/ or www.documentum.com) is a software platform used to manage, store, secure, and publish unstructured documents in a systematic manner, according to predefined business rules, policies, and procedures. EMC offers a Documentum certification program.

Liquent CoreDossier (www.laboratorynetwork.com/ecommcenters/liquent.html) is a software platform for assembling and publishing regulatory documents according to predefined business and regulatory standards.

Regulatory Affairs Professional Organizations

Regulatory Affairs Professional Society (www.raps.org) is a professional organization for regulatory professionals, offering educational, networking, and certification services.

Drug Information Association (www.diahome.org) is a worldwide professional association that fosters innovation and exchange of health information. Members include those involved with discovery, development, regulation, and surveillance of biopharmaceutical products.

Consumer Healthcare Products Association (www.chpa-info.org) is a professional association that promotes leadership and guidance to those who produce over-the-counter drugs and nutritional supplements.

Medical Design and Manufacturing (www.devicelink.com) is an online information source for the medical device industry, providing information, recruiting services, and collaborative opportunities for medical device professionals.

Society of Quality Assurance (www.sqa.org) is an international professional organization promoting regulatory quality assurance of GLP, GCP, and GMP standards worldwide and providing a forum for information exchange and professional development of the quality assurance profession.

Food and Drug Law Institute (www.fdli.org) is a nonprofit organization that provides education and a forum for discussion of legal, policy, and regulatory issues regarding biopharmaceuticals, medical devices, food, and cosmetics.

Profiles of Sponsor CRO Companies (See detailed list in Chapter 13)

Pharmaceutical Research and Manufacturers of America (www.phrma.org) represents the leading pharmaceutical research and biotechnology companies in the United States.

Biotechnology Industry Organization (www.bio.org), the world's largest biotechnology organization, represents more than 1200 biotechnology companies.

Medical Design and Manufacturing (www.devicelink.com) is an online resource for the medical device industry.

Association of Clinical Research Organizations (www.acrohealth.org) represents the world's leading clinical research organizations.

Salary Surveys for Regulatory Affairs Professionals

American Association of Pharmaceutical Sciences Salary Survey (www.aaps.org) is published annually. Access the most recent survey by entering "salary survey" in the website's Search field.

Contract Pharma Salary Survey (www.contractpharma.com) is published annually. Access the most recent survey by entering "salary survey" in the website's Search field.

Medical Device & Diagnostic Industry Salary Survey (www.devicelink.com) is published periodically. Access the most recent survey by entering "salary survey" in the website's Search field and selecting the Research and Development survey.

Salary.com (www.salary.com) publishes salary ranges for positions in a wide range of industries. Enter "regulatory affairs specialist" and if appropriate your target zip code in the Salary Wizard.

9

Entering Clinical Safety

THE CLINICAL SAFETY SPECIALIST IS THE PERSON responsible for collect-ing, coding, organizing, and tracking suspected adverse events that happen to patients during treatment in clinical studies. With any experi-mental medical product, there is always the possibility that patients will experience unanticipated and potentially harmful side effects. Therefore, one critical aspect of conducting clinical studies is ensuring patient safety. In this regard, clinical safety specialists serve two important functions. First, they monitor the data collected from treated patients to spot any sign of adverse reactions to treatment. If so, they work with other medical personnel to ensure that the patient is given the appropriate medical care. Second, they have a mandated regulatory responsibility to report these side effects in a timely manner to regulatory agencies.

Jasmine is the clinical safety specialist assigned to Nancy's clinical study team. Although she officially reports to a supervisor in the clinical safety department, Jasmine works closely with Dr. Abernathy, the CanDo

---

BOX 9-1. *Clinical Safety Specialist Responsibilities*

- Collect, code, organize, and track suspected adverse events
- Review safety section of study protocols, ICFs, CRFs, IBs, and study reports
- Manage SAE assessment, tracking, and reporting
- Maintain adverse event data in a safety database
- Compile adverse events for DMC and safety committee meetings
- Compile annual safety reports
- Train clinical staff on AE reporting procedures

Pharma physician, to assess the medical significance of each adverse event. Jasmine is a CanDo Pharma employee, but the clinical team could alternatively contract the safety assessment services of a contract research organization (CRO).

THE ROLE OF THE CLINICAL SAFETY SPECIALIST

Jasmine's interactions on Nancy's study team are illustrated in Figure 9-1. She does not have supervisory responsibilities, but the clinical safety department provides administrative staff to assist with clerical and administrative tasks. Her purchasing ability is limited by the expense authority level associated with her job title, but the cost of her work in support of the clinical study teams is incorporated into the clinical safety department's annual budget.

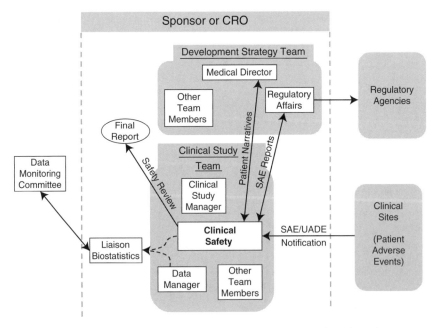

Figure 9-1. Clinical safety specialist interactions with the clinical study team.

Protocol Review

Jasmine begins her work on the study team by reviewing the draft clinical protocol. She specifically looks at the sections dealing with safety assessment. To comply with regulatory requirements, the protocol must include the methods for collecting, assessing, recording, and analyzing adverse events (AEs). In addition, the team must state in the protocol how long and how closely to follow patients who experience an AE. Jasmine's department has developed standard wording for use in the safety section of clinical protocols. This standard text incorporates the regulatory requirements for collecting and reporting AEs, and Jasmine makes only small adjustments so that the text is consistent with the current study's design and objectives.

In conjunction with the protocol, Jasmine also reviews the sample informed consent form (ICF) prepared by the study team and the customized versions prepared by each investigator site. The ICF explains the protocol in easy-to-understand language, including required study visits, procedures, and treatments that will occur if the patient agrees to participate. Because the investigator sites prepare customized versions of the ICF, which are reviewed and may be further modified by their respective Institutional Review Boards (IRBs) before approval, the ICFs for each site in Nancy's study differ. Jasmine checks each customized ICF to confirm that it contains all of the elements required by regulatory authorities regarding patient safety. For example, the ICFs must describe the potential risks and benefits of study participation, instruct patients to report all adverse experiences during treatment, and guarantee that the patient can withdraw from the study at any time.

Case Report Forms and SAE Report Forms

Reviewing and cataloging AE data during Nancy's study is one of Jasmine's most important responsibilities. An AE is any unusual medical condition in a study patient, such as an abnormal laboratory finding, an unfavorable or unintended symptom, or a medical condition temporarily associated with use of the medical product. At CanDo Pharma, the data management and safety departments worked together to compile a standard set of case report form (CRF) pages to gather data on AEs. These standard CRF pages incorporate fields for all of the safety-related information required by regulatory agencies.

In addition to recording these observations on the AE pages of the CRFs, principal investigators (PIs) and sponsors must quickly report adverse events that are classified as "serious" to regulatory agencies and notify the local IRBs. A serious adverse event (SAE) is any unusual medical condition at any treatment dose that results in death, is life threatening, requires hospitalization, or results in persistent or significant disability. (Some medical device studies use the term "unanticipated adverse device effect" [UADE], which is equivalent in concept to the SAE.)

CanDo Pharma's standard SAE reporting form contains fields for all of the information required by regulatory authorities for reporting SAEs. Jasmine makes slight changes to the standard form (e.g., the study number and the experimental drug's name) so that it applies specifically to Nancy's study. Jasmine and Maria, the data manager, include several blank SAE reporting forms in the study binder for each clinical site, along with instructions for notifying CanDo Pharma immediately when an SAE occurs.

Safety Oversight

In addition to Jasmine's safety responsibilities, a data monitoring committee (DMC) will oversee patient safety on Nancy's study. Unlike Jasmine's safety oversight, the DMC has the authority to recommend modifications or termination of the study, based on its independent review of the data. Tom and his biostatistics colleagues are primarily responsible for assisting the DMC, but Jasmine helps the biostatisticians and Dr. Abernathy in designating the DMC membership and preparing for each DMC meeting. (See Chapter 6 for details of DMC activities.)

Sponsors generally establish DMCs only for large, randomized clinical studies that evaluate treatments for major diseases such as cancer or heart disease. For small clinical studies and those that evaluate treatments for less serious medical conditions, CanDo Pharma uses an internal safety committee rather than a DMC.

The internal safety committees operate under CanDo Pharma's standard operating procedures (SOPs) and are managed by the clinical safety department. The internal safety committees consist of CanDo Pharma physicians, biostatisticians, and clinical safety specialists who are not directly involved with conducting the clinical study that they oversee. Based on their review of the safety data, the internal committees, like the DMCs, may recommend whether to continue, modify, or stop the study.

Safety Management Plan

During study start-up, Jasmine also prepares a safety management plan. The plan outlines the procedures that she will use to review, process, and report information on all adverse events collected during Nancy's study. Jasmine reviews her plan with Nancy to ensure that the safety assessment procedures are harmonized with the plans prepared by other study team members such as the data management plan, the monitoring plan, and the statistical analysis plan.

Safety Database

The safety departments at many companies, including CanDo Pharma, prefer to maintain a safety database separate from the clinical study database. At CanDo Pharma, the safety database stores all of the SAE data reported from all of the company's clinical studies. The database also stores side effect information reported by physicians, patients, and pharmacists on CanDo Pharma's marketed products.

Because AE and SAE data are also collected on the CRFs and stored in the study's clinical database, Maria, the data manager, includes a reconciliation plan within her data management plan. Jasmine and Maria agree on how to reconcile the adverse event data in the clinical database, which will be locked at the end of Nancy's study, with the safety department's database, which continues to accumulate SAEs from ongoing and future studies of the drug.

AE Coding

Throughout the study, Jasmine works with the team to clarify AE and SAE information that the PIs report on the AE pages of the CRF. The PIs may use inconsistent descriptions when reporting those medical conditions, but the entries in the clinical study database and safety database must be coded consistently so that the safety risks of the drug can be assessed. Jasmine prepares a "preferred terms" list and uses it to code the reported AEs and SAEs into standard categories. The coding conventions used by most biopharmaceutical companies are defined by the Medical Dictionary for Regulatory Activities (MedDRA).

When the first clinical study on the CanDo Pharma drug was conducted, the safety department established a special section in its safety database for storing SAE data on the drug. Because Jasmine supported the clinical teams that conducted those earlier studies, she has acquired a thorough understanding of the drug's safety profile. She is also familiar with the database procedures for adding new SAE information to the safety database as it is collected during Nancy's study.

Periodically, data specialists in Jasmine's department review the adverse event data. If they find incomplete SAE data or medical descriptions that are inconsistent with Jasmine's preferred terms list, they query the sites for clarification using a process similar to that used by the data management group (see Chapter 5). On the basis of the PIs' clarifications, they assign the appropriate preferred term and apply the appropriate MedDRA codes to ensure consistent classification of the AEs and SAEs.

SAE Reporting

Regulatory authorities require sponsors and PIs to act quickly when a patient experiences an SAE. The clinical site staff and the CanDo Pharma clinical research associate (CRA) immediately notify Jasmine by submitting a completed SAE reporting form. In reviewing the form, Jasmine may ask the clinical site, via the CRA monitor, to provide more details about the patient's condition or the circumstances that led to the SAE. Sometimes, she must use her negotiating skills during these discussions to obtain all of the information that the CanDo Pharma physicians want to see and the regulatory authorities require.

Dr. Abernathy and Nancy's team must accept the PI's classification of the adverse event as "serious" and the PI's decision that the SAE was, or was not, related to drug treatment. However, for PIs and team members who are not familiar with regulatory definitions and SAE reporting procedures, Jasmine patiently guides them through their responsibilities. She often must remind them, for example, of the International Conference on

Harmonization (ICH) criteria for distinguishing between an AE and SAE. This distinction is important, because regulatory agencies scrutinize side effects that are classified as SAEs more closely than AEs.

Jasmine forwards the PI's completed SAE reporting form and the additional medical details to Dr. Abernathy for review. With his input, Jasmine works with Amy, the regulatory affairs specialist, to complete the appropriate reports and inform the regulatory authorities. ICH guidelines and other regulatory agencies state specific deadlines for reporting new SAEs. (See Chapter 8 for details of reporting SAE to regulatory agencies.) In parallel, Dr. Abernathy consults with the PI to determine appropriate medical treatment for the patient.

Jasmine also enters data from the SAE reporting form into the safety database, which is CanDo Pharma's official source of SAE information. As new information becomes available about the SAE, Jasmine updates the safety database and informs Amy, who must update the regulatory authorities. The safety database allows Jasmine to track the status of each SAE, which she follows until the patient's condition is resolved or the medical case is closed.

Data Monitoring Committee

The accumulated AE and SAE data are also reviewed periodically by the data monitoring committee (DMC). Before each DMC meeting, the liaison biostatistician extracts the accumulated AE and SAE data from the clinical database and creates a set of tables, which summarize the types and frequency of reported AEs and SAEs. As a member of Nancy's study team, Jasmine does not know the assignments of patients to the drug and placebo treatment groups. However, she summarizes the blinded AE and SAE data and submits her safety analysis for the DMC's review. She may attend the DMC meetings to discuss the study, but only during the open sessions, at which the DMC only discusses blinded data.

Often, the DMC reviews and makes recommendations based only on the blinded data. Sometimes, however, the DMC may request unblinded data to resolve concerns about patient safety. The liaison statistician, who is isolated from the clinical study, is authorized to unblind the data and conduct further analyses. All discussions of the unblinded data that link the adverse events to specific patients and treatments take place in the DMC's closed sessions, which Jasmine and the other members of Nancy's study team cannot attend. (See Chapter 6 for more information about DMCs.)

Regulatory Reports

Regulatory agencies require sponsors to submit an annual safety report on each of their products, both during development of a new product and after the product is approved for sale. Because Nancy's study continues for several years, Jasmine works with Amy, the regulatory affairs specialist, to prepare and submit a summary of the study's accumulated safety data each year. The annual report lists all AEs and SAEs, grouped by body systems (such as cardiac, gastrointestinal, or genitourinary) and severity (i.e., mild, moderate, or severe). (See Chapter 8 for details of annual reports.)

At the end of the study, Jasmine and Maria reconcile the AE data in the safety and clinical study databases. The SAE entries may differ, because the safety specialists use a different process to resolve safety queries than the data management query resolution process. The reconciliation plan specifies the categories of data that must match in the two databases. These usually include only key features of each SAE such as patient number, clinical site, and onset date.

After database lock, Jasmine works with Tom, the biostatistician, and Dr. Abernathy to summarize the unblinded AE and SAE data and draw conclusions about the CanDo Pharma drug's safety. Although Mike, the medical writer, is primarily responsible for writing the study report, he relies on Jasmine to review all of the AE and SAE tables, graphs, and patient listings for accuracy and completeness. Dr. Abernathy writes the patient narratives for Nancy's study report, but Jasmine is also qualified to write patient narratives. They write narratives for each SAE case and for other patient cases that are particularly noteworthy. (See Chapter 10 for more information about study reports and patient narratives.) Jasmine's signature on the study report approval page indicates her confirmation that the study's safety data are accurate and complete.

Jasmine continues to work with her colleagues in the safety department to gather and summarize safety information, maintain the safety database, and prepare annual safety reports on the CanDo Pharma drug, as long as the company conducts clinical studies and, after approval, sells the drug to patients. Through these activities, Jasmine becomes increasingly knowledgeable about the CanDo Pharma drug's safety profile, and as such, she is frequently consulted by the clinical teams and the regulatory affairs staff.

In addition to her safety assessment duties, Jasmine works with her colleagues in the clinical safety department to conduct training sessions for the clinical teams. Good Clinical Practice (GCP) regulations require rapid and complete disclosure of safety information, and the reporting procedures for AEs and SAEs differ. In her training sessions, Jasmine emphasizes the distinction between AEs and SAEs and the procedures required by regulatory agencies for documenting and reporting SAEs. She also reviews CanDo Pharma's SOPs, so that all team members know their job-specific responsibilities. In addition to Jasmine, other members of the clinical team (e.g., CRAs, study managers, and study physicians) have key responsibilities in processing each SAE.

ONE OF JASMINE'S DAYS

As Jasmine is driving to work, she thinks about her tasks for the day. She has no meetings scheduled, and she plans to continue reviewing ICFs. One of Dr. Abernathy's clinical study teams is starting a large study that involves 150 investigator sites. Each site has modified CanDo Pharma's sample ICF to meet the needs and preferences of their individual IRBs or independent ethics committees (IECs). Copies of those customized ICFs are now arriving, and Jasmine is systematically reviewing them to ensure that the modified text still contains all the elements required by regulatory authorities for protecting patient welfare.

Yesterday, for example, she found that the ICF from one site omitted one of the experimental drug's side effects in the list of foreseeable risks. Another site forgot to include the name and phone number of the person whom patients should contact if they experience an injury or other adverse effects while on the study. She communicated these errors to the CRAs responsible for monitoring those sites and asked them to assist the study coordinators in correcting the ICFs. When she went home yesterday, Jasmine still had a dozen ICFs awaiting her review, and she expects more will be arriving this week.

On her way to her desk, Jasmine, as usual, stops to check the safety department's fax machine for incoming documents. An SAE report from one of the German clinical sites on Nancy's study arrived during the night and is waiting for her. While performing an exercise stress test as required by the protocol, a patient became lightheaded, stumbled, and fell off the treadmill. When he landed, he broke his wrist and bumped his head, causing a gash that required five stitches. The cardiac monitoring data that was collected during the stress test, along with a subsequent examination, revealed that the patient had suffered a mild heart attack. The PI admitted the patient to the hospital for treatment of his injuries and kept him overnight for observation.

At her desk, Jasmine changes her priorities and focuses on her responsibilities regarding the SAE. She carefully reviews the SAE report, which includes all the details required by regulatory agencies, at least for the initial notification. In her e-mail inbox is a message from the PI's study coordinator, alerting Jasmine that the site had faxed the SAE report, in case she missed seeing it. Also, anticipating that Dr. Abernathy and Nancy's team will want additional information, the study coordinator assures Jasmine that the PI will fully cooperate with CanDo Pharma and be available to discuss the case by teleconference.

Jasmine makes several copies of the SAE report and goes to Dr. Abernathy's office to brief him in person. His main concern is whether the patient's dizziness and heart attack were caused by CanDo Pharma's drug. The PI in Germany thinks that those events were related to treatment; however, the study is blinded and the patient might have been given the placebo.

Ultimately, a sponsor must accept the PI's decision regarding the relationship, or lack of it, between an SAE and drug treatment. However, Dr. Abernathy wants more medical information to understand the circumstances surrounding this SAE. He also wants the PI's input on what they should do next: Should the patient continue with the study, assuming that he wants to, or stop treatment? Should he be monitored closely for recurrence of the symptoms? If so, for how long? Jasmine takes detailed notes. By the time they finish reviewing the case, it is too late to reach the German site today. However, Jasmine assures Dr. Abernathy that the PI has already anticipated the need to discuss the case. He asks her to send the PI his list of questions and to schedule the teleconference for tomorrow morning.

Jasmine's next stop is Nancy's desk, where she drops off a copy of the SAE report. Nancy has already left for lunch, but Jasmine leaves a note and mentions tomorrow's teleconference. She knows Nancy will want to attend, and she asks Nancy to invite other members of her study team, as appropriate.

After lunch, Jasmine goes to Amy's office in the regulatory affairs department. She gives Amy a copy of the SAE report, briefs her on the discussions with Dr. Abernathy, and informs her about tomorrow's teleconference. To comply with ICH requirements, Amy must submit details of the SAE to the regulatory authorities in all of the countries where CanDo Pharma is conducting studies of its experimental drug within 15 days of the time stamped on Jasmine's incoming fax. They briefly discuss their plans for compiling the appropriate regulatory information. Although the PI's SAE report is sufficient for the initial regulatory notification, Amy wants to wait for the outcome of tomorrow's teleconference. Everyone will then have a better understanding of the SAE, the patient's condition, the plans for his continued treatment and medical oversight, and whether to break the blind and disclose the patient's treatment assignment.

CanDo Pharma has a branch office in France but not in Germany. Amy gives Jasmine the name of her regulatory affairs counterpart in CanDo Pharma's French office and asks her to include him in the teleconference. He will assist Amy in submitting the appropriate SAE reports to the regulatory authorities in Europe, including most importantly, Germany.

When Jasmine returns to her desk, she makes the arrangements for the teleconference. In addition to herself, the participants will include Dr. Abernathy, Nancy, Amy, the French regulatory affairs specialist, the German PI and study coordinator, CanDo Pharma's German-based CRA, and selected members of Nancy's study team. The agenda for the teleconference primarily consists of the list of questions that Dr. Abernathy wants to address with the German PI regarding the patient's pre-existing medical condition, the circumstances surrounding the SAE, the patient's follow-up treatment and prognosis, and whether to continue the patient on the study.

Dr. Abernathy will not attempt to influence the PI's decision about whether the SAE was caused by the CanDo Pharma drug; however, he wants to make sure that the PI considers all of the facts and makes an appropriately informed decision. Dr. Abernathy also wants to fully understand the SAE. If it is justified, he will amend the treatment protocol to avoid occurrence of the same situation in other patients.

After Jasmine e-mails the teleconference details and agenda to all of the participants, she returns to her review of the ICFs. Nancy stops by her desk and asks Jasmine to summarize Dr. Abernathy's assessment of the SAE. They know that no other patients in the study have suffered a heart attack, but some patients have experienced dizziness or headaches. Nancy wonders whether dizziness could be an early warning sign leading to heart problems in the study patients. Although these side effects probably did not cause the heart attack, she asks if Jasmine could review the safety database before tomorrow's teleconference and bring a summary of the patients who have reported dizziness or headaches.

Jasmine explains that those symptoms are frequently reported by study patients, whether they receive drug treatment or not, and the safety database does not routinely track them. The clinical study database, on the other hand, tracks all of the study patients' side effects, no matter how small. Maria, the data manager, would be able to extract that information from her database. Nancy asks Jasmine to include both Maria and Tom, the study's biostatistician, in tomorrow's teleconference. If Dr. Abernathy thinks additional information would be helpful after speaking to the German PI, they will be fully informed and know which data points to extract for closer scrutiny.

After Nancy leaves, Jasmine finishes her preparations for the teleconference and invites Maria and Tom. It is now too late to review the remaining ICFs on her desk. If things go smoothly at tomorrow's teleconference, she can then return to her ICF reviews. More likely, though, the teleconference will generate a series of action items, some of which she will be responsible for handling. And, those actions will probably need to be completed before Amy files her mandatory regulatory notifications 2 weeks from now.

HOW JASMINE GOT HER SAFETY SPECIALIST JOB

Requirements

Jasmine has a bachelor's degree in nursing and began her career as a practicing nurse at a major hospital center. Assessing product safety and processing SAE reports requires a good understanding of human health and disease. Therefore, most clinical safety specialists are licensed nurses.

After working several years as a nurse, Jasmine accepted a position with a health insurance company, where she learned to code medical claims for health benefits. She learned how to classify medical conditions, and the experience reinforced her understanding of diseases and medical treatment.

BOX 9-2. *Requirements for a Clinical Safety Specialist Position*

- B.S. in nursing, registered nurse licensure, or equivalent
- Clinical experience: 2 to 4 years of nursing experience or equivalent
- Knowledge of medical and drug terminology
- Familiarity of GCPs
- Familiarity with worldwide regulatory requirements and adverse event reporting
- Knowledge of clinical databases
- Strong computer skills (MS Office)
- Excellent written and verbal communication skills
- Anticipate and identify problems and take appropriate action
- Excellent organization skills
- Attention to detail and high degree of accuracy
- Good team player
- Ability to work independently

Although Jasmine enjoyed her work at the insurance company, she realized that many diseases do not have adequate treatment options, and she wanted to be more involved with developing new and more effective therapies. She realized that she could use her nursing and insurance claims knowledge in the pharmaceutical industry as a clinical safety specialist.

Finding a Clinical Safety Specialist Position

You can find entry-level clinical safety positions at large sponsor companies and at large CROs. Small companies typically hire a consultant to handle their safety assessment work. Most companies post clinical safety job openings on their websites, which can easily be searched by job title. The job titles for entry-level clinical safety positions vary considerably between companies, but look for titles such as clinical safety specialist, drug safety associate, medical device safety associate, or safety support specialist.

Large sponsor and CRO companies typically have fully developed clinical safety departments. The experienced staff, established procedures, and validated computer systems offer an entry-level worker the best opportunities for gaining high-quality training, experience, and career advancement. Some companies maintain one safety function to support clinical studies of experimental products (often called clinical safety) and another to support products after marketing approval (often called pharmacovigilance). Other companies combine clinical safety and pharmacovigilance into one safety department or may incorporate safety as a unit within the regulatory affairs department. (See Chapter 8 for information about regulatory reporting of safety data.)

The hiring managers at both sponsor companies and CROs look for the same skills and qualities in their job candidates: knowledge of human physiology in health and disease, medical recordkeeping, and communication skills. Jasmine's training and experience as a nurse gave her the necessary qualifications for an entry-level clinical safety position, but there are other ways. Some nurses gain relevant experience by processing insurance claims for doctors in private practice. Others become familiar with recordkeeping through their administrative activities in a teaching hospital or a nonprofit organization such as the Red Cross.

Nurses may also enter clinical safety jobs after working as a study coordinator or in clinical data management. Study coordinators gain important experience in conducting clinical studies and reporting AEs

---

BOX 9-3. *Tips for Getting a Clinical Safety Specialist Position*

- Experience as a practicing nurse.
- Experience with medical recordkeeping
- Consider learning medical claims coding for insurance companies or practicing physicians
- Consider administrative positions at a teaching hospital or the Red Cross
- Consider working as a study coordinator or in clinical data management
- Take online courses on GCP
- Join a professional society such as DIA
- Search clinical safety job postings on websites of drug, medical device, and CRO companies
- Preferentially look for entry-level jobs at large CROs and biopharmaceutical companies

and SAEs. (Chapter 3 gives details for landing a study coordinator position.) Nurses with data management experience are familiar with entering, coding, and reviewing clinical data, as well as GCP and other regulatory requirements. (Chapter 5 gives details for entering data management positions.)

Landing the Job

All of the hiring managers ranked Jasmine as a highly desirable clinical safety specialist candidate. Although she had never used the MedDRA coding system, she was familiar with the standard medical coding schemes required by her health insurance employer. She understood medical and drug terminology, and she knew the importance of entering accurate medical information into the medical claims database. Furthermore, her supervisors and coworkers confirmed that she worked efficiently and always met her deadlines.

The hiring managers were less impressed with the candidates who competed with Jasmine for entry-level positions, even though some of them were study coordinators with a degree in nursing. Clinical safety specialists must collect and communicate adverse event data accurately, and they must work diligently to document the resolution and closure of SAE incidents. The hiring managers therefore rejected candidates who could not provide examples of their problem-solving skills and their ability to complete assigned tasks successfully. Hiring managers were also concerned about candidates who had poor interpersonal skills. Clinical safety specialists must be firm but tactful when discussing SAE details with the study team and physicians.

Working for a Sponsor Company versus a CRO

Although Jasmine could have taken an entry-level clinical safety position at a large CRO, she decided to accept an offer from CanDo Pharma. Because she had not worked as a study coordinator and was not familiar with the operational aspects of clinical studies, her first assignments were mainly administrative: routing incoming faxes of SAE reports, entering data into the department's safety database, and maintaining the central safety files of CanDo Pharma's products. In parallel, Jasmine successfully completed CanDo Pharma's required training modules, including GCP training, the safety department's SOPs, and MedDRA coding.

As Jasmine gained on-the-job experience and familiarity with regulatory requirements, she took more important assignments. Under the guidance of experienced safety specialists, she reviewed and provided feedback on the safety sections of study protocols, CRFs, investigator's brochures (IBs), and ICFs. She coded adverse events using the MedDRA codes. Eventually, her supervisor allowed her to prepare official safety reports and draft patient narratives. (See Chapter 10 for more information about patient narratives.)

To supplement her knowledge of regulatory and safety procedures, Jasmine joined a professional society. (The Drug Information Association [DIA] and the International Society for Pharmacoepidemiology [ISPE] are two prominent societies for safety specialists.) The advantages of membership include training classes, an official safety and pharmacovigilance certification program, newsletters and other clinical safety-oriented publications, and an annual society meeting.

Jasmine's career advancement in clinical safety depends heavily on her reputation as a clinical safety specialist, and she can build her reputation most efficiently through her performance and networking. Clinical safety specialists earn the respect of their teams and the company's physicians by accurately assessing medical conditions, being diligent and timely in compiling safety information, and assisting the teams and regulatory specialists in submitting required regulatory notifications and reports. Clinical workers value knowledgeable clinical safety specialists who help them comply with safety requirements and are likely to recommend them for advancement.

Because Jasmine serves as the clinical safety representative on many study teams, she meets many physicians and managers in the clinical department. If she impresses them with her safety knowledge and professionalism, they can provide influential references and recommend her for promotion. Jasmine's listing in her professional society's membership directory also makes her visible to recruiters and hiring managers who are looking for experienced clinical safety specialists.

By the time Nancy's study started, Jasmine had served as the clinical safety specialist on several other clinical study teams, including previous studies in Dr. Abernathy's development program. She was familiar with the drug's safety profile and was fully prepared to oversee the safety reporting activities on Nancy's study.

MOVING FORWARD

Experienced clinical safety specialists are in high demand and generally receive salaries that are higher than those paid to nurses working in hospitals or other healthcare settings. Many clinical safety specialists, like Jasmine, decide to continue their careers within clinical safety and gain more experience in handling AEs and SAEs. Although the titles of clinical safety positions (e.g., drug safety specialist, pharmacovigilance associate, medical device safety associate) vary considerably, large sponsor companies and large CROs offer career tracks in drug safety and medical device safety. From an entry-level position, clinical safety workers can expect promotions every 3–5 years to the next level of responsibility, depending on performance. Typically, clinical safety specialists like Jasmine reach a manager-level position after gaining experience at two or three lower levels.

Clinical safety managers supervise junior-level staff and take greater responsibility for handling adverse event data. They train new employees on departmental and regulatory procedures, determine the safety specialists' work assignments, and manage the standards for quality of the safety database. When difficult or unusual SAEs occur, they may choose to handle the safety reporting procedures personally.

Clinical safety managers who have medical credentials may advance to senior positions in a clinical safety department such as safety medical officer or department head. Safety officers are often responsible for global safety oversight of a single medical product or specialized safety oversight of a group of products. In the latter case, the products may be grouped by stage of development; for example, one safety officer usually oversees the adverse events from all products during Phase 1 (or Pilot) clinical studies. Alternatively, products may be grouped by therapeutic area; for example, one safety officer may oversee all cancer drugs and another oversees cardiology products.

Safety officers periodically review the accumulated data in the safety database for each product, looking for trends and patterns of AEs. Even when individual observations are not troublesome for individual patients, the AEs may show a trend or pattern when the observations from many patients are combined. The accumulated trends and patterns may reveal situations (such as pre-existing medical conditions or specific patient characteristics) under which the product is more likely to produce AEs.

Sometimes, these trends and patterns raise concerns about the product's overall safety. The safety officer interprets the trends and patterns and makes recommendations to the strategy team for further evaluation and follow-up.

One of the most important responsibilities of a global safety officer comes at the end of the development program. When Dr. Abernathy and his strategy team have completed all of the clinical studies of the CanDo Pharma drug, they compile the data in a marketing authorization application (MAA). The global safety officer assists the medical writer and the biostatistician in compiling the safety data and writing the clinical summary of safety (CSS) section of the MAA. The CSS summarizes data relevant to safety in the intended patient population, integrating the results of individual clinical study reports and the safety analyses contained in previously submitted annual reports. From this information, the global safety officer and the team construct the drug's safety profile, which is summarized in tables and graphs in the MAA. (See Chapter 8 for details of MAA preparation and submission to regulatory authorities.)

Instead of concentrating on experimental products, Jasmine could choose to work in the pharmacovigilance group, which monitors the safety of CanDo Pharma's marketed products. Although CanDo Pharma conducts a few, specialized clinical studies after its products are approved for sale, most of the side effect data on marketed products come from unsolicited reports by physicians, pharmacists, patients, and the families of patients. The pharmacovigilance group collects, reviews, and adds this information to the safety database. Each year, CanDo Pharma must report a summary of the product's updated safety profile to the regulatory authorities. If the accumulated data indicate new concerns about the product's safety, the regulatory reviewers may require CanDo Pharma to restrict the product's use, conduct new studies to better understand the product's side effects, or perhaps withdraw it from sale altogether.

Alternatively, Jasmine may wish to pursue a different career path such as medical information. The scope of work conducted by medical information departments varies, but in general, medical information workers answer external and internal questions about their company's products. External inquiries come from healthcare professionals, patients, and the general public. Internal inquiries come from the company's management and staff. To answer those questions, the medical information workers must remain current in their medical knowledge about their company's products, as well as similar products marketed by the company's competitors. (See Chapter 10 for more details about medical information.)

Jasmine could also use her knowledge of clinical safety to move into a regulatory affairs position or work as a project manager. Regulatory affairs and project managers have a more strategic and highly visible role in product development. Their responsibilities allow them to interact with people on the multidisciplinary clinical teams, as well as the company's senior executives. They help team leaders to harmonize individual product development activities, ensuring that all regulatory requirements are met efficiently and with the greatest probability of success. Chapter 8 describes regulatory affairs positions and Chapter 11 discusses project manager positions in greater detail.

CLINICAL SAFETY SPECIALIST RESOURCES

Good Clinical Practices Training and Certificate Programs

Drug Information Association (www.diahome.org) offers a global certificate program for clinical safety and pharmacovigilance.

ClinfoSource (www.clinfosource.com) offers online GCP courses and certification for CME and CNE credit at reasonable cost.

Kriger Research Center International (www.krigerinternational.com) offers online GCP courses and certification in multiple languages on the country-specific web pages under the "Courses" tab.

Association of Clinical Research Professionals (www.acrpnet.org) offers online GCP courses and certification for CME, CNE, and ACRP credit at reasonable cost.

GCP Training Online (www.gcptraining.org.uk) offers online GCP training and certification, emphasizing the European Clinical Trial Directive, at reasonable cost.

Medical Coding Dictionaries

Medical Dictionary for Regulatory Activities (www.meddramsso.com) is a clinically validated dictionary of terms used to report adverse event data from clinical studies.

Clinical Safety Professional Organizations

International Society for Pharmacoepidemiology (www.pharmacoepi.org) is an international organization providing a forum for exchange of scientific information, education of, and advocacy for, the field of pharmacoepidemiology and pharmacovigilance.

Drug Information Association (www.diahome.org) is an international professional association for those who are involved in discovery, development, regulation, surveillance, or marketing of biopharmaceutical products.

Profiles of Sponsor and CRO Companies (*See detailed list in Chapter 13*)

Pharmaceutical Research and Manufacturers of America (www.phrma.org) represents the leading pharmaceutical research and biotechnology companies in the United States.

Biotechnology Industry Organization (www.bio.org), the world's largest biotechnology organization, represents more than 1200 biotechnology companies.

Medical Design & Manufacturing (www.devicelink.com) is an online resource for the medical device industry.

Association of Clinical Research Organizations (www.acrohealth.org) represents the world's leading clinical research organizations.

Salary Surveys for Clinical Safety

Salary.com (www.salary.com) publishes salary ranges for positions in a wide range of industries. Enter "drug safety" and if appropriate your targeted zip code in the Salary Wizard.

10

Entering as a Medical Writer

M EDICAL WRITING IS A PROFESSION that serves many healthcare organizations. In addition to companies that produce drugs and medical devices, these include university medical centers, hospitals, government agencies, and advertising and marketing agencies.

Medical writers who work in the biopharmaceutical and medical device industry are responsible for writing documents such as protocols, clinical study reports, investigator's brochures, product monographs, marketing brochures, and promotional materials. They also assist scientists and clinicians in writing abstracts for presentation at medical conferences and manuscripts for publication in medical journals.

Sponsors must communicate volumes of information to regulatory agencies, and they recognize the public relations value of distributing high-quality information about their products to the general public. Medical writers who can write highly technical regulatory documents and succinctly worded promotional materials are essential in meeting these regulatory and public relations obligations.

Some medical writers, like Mike, work in the clinical or regulatory department and write scientific and technical documents for professional audiences. Other medical writers support corporate communications, sales and marketing activities, and write for consumers and lay audiences as well as medical professionals. Mike is a CanDo Pharma employee, but CanDo Pharma occasionally contracts medical writers from contract research organizations (CROs) and freelance medical writers to write documents on its behalf.

Medical writers typically do not have supervisory responsibilities, but Mike can rely on clerical and administrative support from his department. The head of Mike's department prepares and maintains an annual budget to cover departmental expenses, including those in support of clinical studies. Typically, these expenses are modest and mainly consist of

BOX 10-1. *Medical Writer Responsibilities (in the biopharmaceutical industry)*

- Write clinical protocols, sample ICFs, IBs, and clinical study reports
- Write annual reports, summaries, and applications for regulatory submission
- Lead document review meetings
- Write medical education materials, training aids, and patient information sheets
- Write clinical study and results entries for public-access websites
- Write medical device Instructions for Use
- Draft physician and patient information sheets
- Prepare slide presentations and posters
- Write white papers and company position papers
- Write pharmaceutical marketing and advertising copy
- Participate in document quality control
- Edit and proofread scientific and medical manuscripts authored by others
- Assist scientists and clinicians in writing abstracts and journal articles

the cost of purchasing and maintaining word processing and document management software and hardware. Mike's purchasing ability is limited by the expense authority level associated with his job title. However, he can easily receive authorization to negotiate a contract with a CRO or freelance medical writer if his department decides to outsource medical writing assignments. In those cases, Mike would oversee the work of the external medical writers and ensure that they meet CanDo Pharma's document requirements.

THE ROLE OF THE MEDICAL WRITER

Mike began working with Dr. Abernathy and his strategy team when they were planning the development program on the CanDo Pharma drug. With input from the team, Mike wrote the investigator's brochure (IB) and worked with Amy, the regulatory affairs specialist, to write the documents required by the regulatory agencies for the Investigational New Drug Application (IND) and Clinical Trial Authorization (CTA) applications.

Templates and Style Manuals

While at CanDo Pharma, Mike has written many regulatory documents, including numerous protocols, study reports, IBs, and annual reports for several experimental drugs and several IND/CTA applications. In addition, he has worked with other medical writers to write the integrated safety and efficacy summaries for marketing authorization applications (MAAs), which CanDo Pharma must submit to regulatory authorities for market approval of each new drug. (See Chapters 2 and 8 for more information about regulatory documents.)

All of the documents that CanDo Pharma submits to regulatory authorities are written according to corporate standards for style and format. Mike and his colleagues worked in conjunction with the other clinical departments to develop those standard templates, which were based on regulatory agency requirements. Each template sets the default document format (e.g., headers, footers, page numbering, section headings, margins, font styles, and font sizes), provides standard text for some sections of the document, and gives instructions or examples on how to complete other sections. These text and formatting standards help readers locate key information, because all CanDo Pharma reports are organized identically. The standards also streamline the filing system in CanDo Pharma's document archives. Most importantly, the standard formats make it easy to merge all of the clinical reports and other regulatory documents into the comprehensive MAA for regulatory review.

Writing Regulatory Documents

When Dr. Abernathy and the team begin planning Nancy's clinical study, Mike had already written reports and regulatory documents for earlier studies in the development program and is therefore familiar with the CanDo Pharma drug. He begins his work on Nancy's study by drafting the study protocol. Some sections of the protocol are provided by the team's "specialists," such as the statistical rationale from Tom, the biostatistician, and the adverse event reporting procedures from Jasmine, the clinical safety specialist. In combining this material with his text, Mike follows CanDo Pharma's document style and formatting standards. While he is writing the draft and during protocol review, Mike also serves as a proxy for external readers and makes sure that people unfamiliar with the drug and the development program will be able to understand the protocol and its procedures.

After the protocol is finalized and while the study is ongoing, Mike continues to attend team meetings and supports Nancy and her colleagues by writing various regulatory documents. For example, each year, Mike writes the annual report, which updates the regulatory agencies on the status of the CanDo Pharma drug. Biostatisticians extract the accumulated data from all completed and ongoing studies of the drug, including Nancy's study, and summarize them in tables and patient listings. Clinical safety specialists focus on the adverse event data, assist in interpreting the safety data, and note any trends or changes in the drug's safety profile. Under the direction of the regulatory affairs department, Mike combines all of this information and writes the annual safety report. (See Chapters 8 and 9 for further details about annual safety report preparation.)

Regulatory authorities require sponsors to update their IBs when significant new data become available. For each update, Mike works with Dr. Abernathy and other members of the strategy team to interpret the new data and write the update using CanDo Pharma's official IB format. He coordinates his work with Amy, the regulatory affairs specialist, who submits the IB update to each regulatory agency. In conjunction with the clinical study managers, including Nancy, he ensures that the updated IB is also distributed to each principal investigator (PI).

Clinical Study Reports

At the end of Nancy's study, Mike takes the data tables, patient listings, and the biostatistician's conclusions; interprets the data; and writes the clinical study report. Even during the early drafts of the clinical report, Mike works closely with the other members of Nancy's team and with Dr. Abernathy. They provide guidance and answer his questions about the study. Together, they ensure that the report states the study's objectives, procedures, and results clearly and accurately. Sometimes, Mike must referee conflicting written comments or spirited discussions among team members, to decide how to interpret the results and express the study's conclusions in the report.

Most clinical study reports also include patient narratives. A patient narrative describes an individual patient's medical condition and response to treatment without disclosing the patient's identity. Narratives are typically written for patients whose disease state or response to treatment is unusual or particularly noteworthy. Mike must also include patient narratives for each serious adverse event (SAE) that occurred during the study. The SAE narratives describe the nature of the SAE, the clinical course leading up to the SAE, countermeasures that were taken, whether experimental drug treatment was stopped, and an interpretation of each patient's medical condition. Either Dr. Abernathy, the study physician, or a nurse in the clinical safety unit usually writes the patient narratives, but sometimes they give their notes to Mike and ask him to write them.

Typically, clinical study reports go through two formal approval stages. After circulating several drafts of the report to Nancy's team for comments and review, Mike submits the report one final time to key team members for formal approval. If they spot errors or disagree with the way the report presents the study's results and conclusions, they will ask Mike to make changes before they sign the report's approval page. Their signatures indicate that the report is accurate and that it represents the team's best work.

In the second stage of approval, Mike submits the report to Dr. Abernathy, key members of the program strategy team, and, importantly, CanDo Pharma's senior management. The second-stage reviewers may ask for clarification or changes that require Mike to hold further discussions with the study team. After the second-stage reviewers sign the report's approval page, the clinical study report is considered final.

Mike and Nancy, the study manager, submit the final study report to CanDo Pharma's document archive. Once in the archive, Mike notifies Amy, the regulatory affairs specialist, and the other CanDo Pharma managers that the clinical study report is available.

In conjunction with the study report, Mike also prepares a short summary (or synopsis) of the study. This summary includes the study plan, objectives, inclusion criteria, methodology, and results. It provides enough detail to serve as a standalone document and has several purposes. Many people at CanDo Pharma use the summary to communicate the highlights of the study internally. In addition, David, the clinical research associate (CRA), sends the summary to the PIs who participated in the study. Because the PIs each enrolled and treated only a portion of the study patients, each of whom was blinded, the PIs would not otherwise know the overall outcome of the study. Some PIs must submit this study summary to their respective institutional review boards (IRBs) as part of their required close-out procedures. Finally, Nancy and Amy may ask Mike to post the summary on a public-access website such as www.clinicalstudyresults.org so that the results are available to anyone who is interested in the study's outcome.

Mike is a member of several clinical study teams and spends most of his time writing clinical study reports. Because the clinical studies are at different stages, each taking several years to complete, he can easily shift from one study to another and write the corresponding reports and other regulatory documents as they are needed.

Writing Nonregulatory Documents

In addition to clinical study reports and other regulatory documents, CanDo Pharma also distributes materials that highlight the company's scientific accomplishments and support its marketing efforts. These include high-profile scientific publications, communications at scientific and medical conferences, product monographs, training manuals, newsletters, and marketing materials. Some of these documents are written by Mike and his colleagues in the regulatory medical writing group; others are written by medical information or marketing medical writers.

For product monographs (or dossiers), the intended audience determines which medical writing group takes charge. A product monograph for a marketed product summarizes the properties, claims, indications, and conditions of use for a drug in a factual, straightforward manner. In most countries, medical information writers prepare the product monograph, which is a key reference document for managed-care organizations and other healthcare providers and payers.

Some countries (for example, Canada) require a product monograph as part of the official regulatory documentation, and it includes specific instructions on the optimal, safe, and effective use of the drug. In some cases, the product monograph serves as the regulatory equivalent of the IND in the United States. Mike would likely write those product monographs in conjunction with a regulatory affairs manager. For medical devices, medical writers like Mike write the Instructions for Use to provide details for the product's label or an accompanying leaflet regarding the safe and proper use of the device.

CanDo Pharma's clinicians and their PI collaborators gain professional recognition by publishing papers in prestigious medical journals. Mike may serve as the clinical authors' editor, revising and improving their drafts. At other times, Mike assists scientists and clinicians in developing their original drafts as well as providing editorial support; on these manuscripts, Mike may qualify as a coauthor.

Mike also writes materials for medical conferences. The national and international conferences sponsored by medical and scientific societies offer a forum for disseminating information that showcases CanDo Pharma's products and the team's scientific accomplishments. Medical writers such as Mike help the scientists and clinicians to write the abstracts, posters, and visual aids for their presentations. In parallel, marketing medical writers prepare a variety of publicity materials that are timed to coincide with the meeting. These include website postings, press releases, product brochures, and display materials for the company's booth in the conference exhibit hall—all featuring CanDo Pharma's products.

Medical Writing for Marketing and Medical Information

The medical writers in the marketing and medical information departments help CanDo Pharma's marketing staff and sales representatives by compiling materials that will raise awareness of CanDo Pharma's products to the general public. They write some materials for healthcare professionals and other materials for the news media and nontechnical readers. Examples include:

- Educational materials for sales representatives, healthcare professionals, or lay audiences describing a specific disease or health problem

- Training manuals describing how to administer a drug or use a medical device

- Promotional and corporate communications materials that highlight a new drug or medical device's advantages

- Press releases that announce a new medical breakthrough at the company

The written materials produced by medical information and marketing writers take a wide variety of forms, depending on their intended use. Traditionally, these writers prepared printed pamphlets, brochures, and posters, which sales and marketing representatives distributed to prescribing physicians for placement in their waiting rooms or posting in their examination rooms.

Increasingly, however, medical writers produce materials in electronic formats, which offer many advantages. Medical information writers can post and update product information quickly and efficiently on the company's website. Regulatory medical writers like Mike post information about ongoing and completed clinical studies on public-access websites. Through those websites, patients and other readers can contact sponsor companies directly, ask questions, and request specific information about the company's studies and products. CD-ROMs and DVDs of educational and training information may be interactive and may include animation that is not possible with printed materials. In all of these ways, medical writers support the company's efforts to maintain its visibility to patients and healthcare professionals, inform the general public, and promote its products.

Figure 10-1 summarizes the types of medical writers, the respective groups they serve, and the documents they produce.

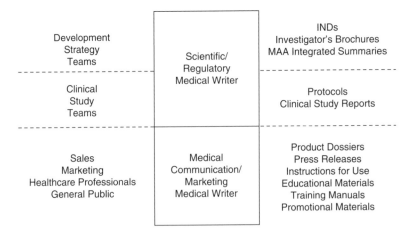

Figure 10-1. Types of medical writers, the groups they serve, and the documents they produce.

ONE OF MIKE'S DAYS

When Mike arrives at his desk today, he plans to continue working on three clinical study reports that are at different stages of completion. He is facilitating the signature approvals on one report that is now final, writing the first draft of a second report for which data analysis is now complete, and preparing to write the report for Nancy's study on which the biostatistician is now doing his preliminary analysis.

The first of these reports is now in the second stage of the review and approval process. During the first stage, the study team had disagreed on how to interpret some of the results and requested a number of text changes, some of which conflicted with Mike's interpretation. Mike facilitated discussions with the team members, and they finally agreed on changes that still reflected the study's major findings, some of which were more impressive than others. Now that these contentious issues have been resolved, Mike does not anticipate any further modifications to the text. However, key people in CanDo Pharma's senior management must review and sign the report before he can submit it to the archive.

He also knows that Amy, the regulatory affairs specialist, is waiting to submit the final, signed report to the regulatory agencies. She submitted his draft report several months ago, so that CanDo Pharma could inform the authorities as soon as possible about the study's key results. Although the authorities appreciated CanDo Pharma's willingness to disclose the results quickly, they nevertheless only accept final study reports as official, and Amy is anxious to submit it.

Mike's responsibility is to keep track of the report as it reaches each senior manager's desk and make sure the report is signed without delays. He knows that CanDo Pharma's executives are easily distracted by more interesting activities, and reports languish unsigned for a long time unless he sends gentle reminders. This morning, he contacts the two executives who have not yet signed the report, points out the importance of submitting it to the regulatory agencies, and asks them to notify him as soon as they have signed it.

Mike spends the rest of the morning continuing to write the draft of the second study report. This was a small study that examined whether diet affected the therapeutic effects of one of CanDo Pharma's experimental drugs. The data confirmed that the drug worked equally well, whether the patients took the drug on an empty stomach, after a fatty meal, or in conjunction with different beverages. Because the results do not restrict or alter the company's plans for treating patients and certainly do not indicate any safety risks, the urgency in completing this report is far less than the one on which Amy is waiting. Mike's progress on the diet study report has already been delayed a number of times by higher priority tasks. Nevertheless, the CanDo Pharma standard operating procedures (SOPs) and Good Clinical Practice (GCP) regulations both require him to produce a final study report, and he would like to check it off his To Do list.

Mike also has been preparing to start writing the study report for Nancy's study. He knows the data have been unblinded and Tom, the biostatistician, has been doing his preliminary analysis. Even in the absence of the analyzed results, Mike can write some sections of the final report, such as the introduction, study objectives, study design, patient selection criteria, and treatment methods. He has started writing those sections, but Tom's full data analysis will take several weeks. So, Mike hopes that he will have a few quiet days to finish the draft of the diet study report and move it off his desk before returning to his incomplete draft of Nancy's study report.

When Mike returns from lunch, a voice message from an excited Dr. Abernathy is waiting for him. Mike returns the call and learns that Dr. Abernathy has had a busy morning. The strategy team met with Tom to review his just-completed preliminary analysis of Nancy's study. The results were more impressive than anyone had expected, and the team wants to submit a paper for presentation at an upcoming international medical conference. Dr. Abernathy managed to obtain authorization from CanDo Pharma's executives to disclose the results at the conference, but only if the abstract submission is coordinated with a press release by the company's public relations department.

Because Mike has been serving on Nancy's study team and is the medical writer most familiar with the study, Dr. Abernathy asks him to write the abstract with Tom's help and cooperate with the public relations officer, who will prepare the press release. The conference's abstract submission deadline is tomorrow, but before then, CanDo Pharma's executives want to see and approve the abstract. The press release must also be approved and ready to send to the media at the same time as the abstract is submitted to the conference organizers.

Mike puts aside his other reports and reads the guidelines for submitting abstracts to the conference. For the next hour, Tom briefs him on the preliminary results, and they agree on the study highlights that should be included in the abstract. After Tom leaves, Mike writes a rough draft of the abstract and e-mails it to Dr. Abernathy and Nancy for their comments. They both reply quickly, and Mike is reading their suggested changes when the public relations officer arrives at his desk to discuss the press release.

She has already gathered material that she will craft into quotes from Dr. Abernathy and the CanDo Pharma executives regarding the significance of the results. Mike shows her his draft abstract, and they discuss how to present the study highlights so that the two documents are not contradictory. She wants to use some words in the press release that Mike feels will distort the actual findings of the study, and he gently but firmly offers some alternatives.

They are just completing their revisions of the abstract and press release when Dr. Abernathy arrives, reads the text of both documents, and gives his approval. Tomorrow morning, he will hand-deliver the abstract to the CanDo Pharma executives for their approval. In parallel, the public relations officer will obtain approval of the press release. They both ask Mike to stand by in case any last minute changes are needed on either document.

Hopefully, the edits, if any, will be minor, and he can then return to his other study reports. At least now he knows the general outcome of Nancy's study, and he will think about how to write her study report with those results in mind.

HOW MIKE GOT HIS MEDICAL WRITING JOB

Requirements

Mike received his bachelor's degree in biology and decided to continue his studies in life sciences in graduate school. In addition to his scientific coursework, Mike always enjoyed writing. As a graduate student, he coauthored several journal articles describing his laboratory work. His first major writing product, however, was his Ph.D. thesis. Luckily, Mike's thesis advisor generously devoted her time to instruct him on scientific writing. Mike also assisted his advisor in writing grant proposals for research funding and wrote articles for the department's newsletter.

Mike's postdoctoral fellowship broadened his writing experience. In addition to reporting his laboratory studies, he took courses to learn

---

BOX 10-2. *Requirements for a Medical Writing Position*

- B.S., M.S., Ph.D., or Pharm.D. in life science or related health sciences
- Clinical experience: 1–3 years experience writing pharmaceutical or health-related documents
- Ability to interpret and analyze scientific and medical data
- Excellent grammatical and communication skills, both written and oral
- Excellent interpersonal and presentation skills
- Ability to write clearly and concisely in scientific document format
- Advanced word processing skills and proficient with electronic templates
- Proficiency with job-related software (MS Word, Excel, PowerPoint, Visio, Adobe Acrobat, Illustrator)
- Strong organizational skills
- Ability to manage and prioritize multiple tasks in a fast-paced setting
- Attention to detail, accuracy, and legibility
- Ability to work with complex projects and within cross-functional teams
- Knowledge of regulatory document requirements, GCPs, and ICH guidelines
- Familiarity with quality assurance and quality control procedures

---

medical terminology. His thesis and postdoctoral work also gave Mike experience in organizing scientific data and drawing conclusions from the results.

An American Medical Writers Association (AMWA) survey indicates that about one-third of medical writers have doctoral-level degrees. The remaining medical writers are evenly divided between those with master's degrees and those with bachelor's degrees. Most medical writers hold degrees in science or healthcare; approximately one-fourth of all medical writers earned degrees in the humanities.

Many colleges and universities offer courses and certificate programs in medical writing. Students in these programs learn scientific and technical writing through structured assignments that help them build a portfolio, which they can show prospective employers.

Mike started his career as a research scientist at a mid-sized biotechnology company. He wrote and published journal articles describing his research findings. In addition, he volunteered to write the nonclinical sections of several IND applications that his company was preparing. Mike wrote his assigned regulatory documents under the direction of a regulatory affairs manager, who was responsible for compiling the IND applications. In addition to his writing assignments, Mike closely followed the discussions of the clinical teams as they prepared their clinical protocols and the clinical sections of the IND. Through these activities, Mike gained a clear understanding of how to write regulatory and clinical documents.

Mike enjoyed laboratory work, but he found more satisfaction in organizing the data, describing the methods and results, and explaining their significance. Having seen his colleagues in action as they produced regulatory and clinical documents, he realized that he wanted to work as a medical writer full-time. As a medical writer, he would continue to use his scientific knowledge and be close to new scientific breakthroughs, but he could devote all of his time to interpreting and communicating those findings in published, or regulatory-sanctioned, documents. He liked the idea of working quietly at his desk, while still having close collaborations with other healthcare professionals through his writing assignments.

Finding a Medical Writing Position

You can find entry-level medical writer positions at large and mid-sized CROs and sponsor companies. Small companies typically do not use full-time medical writers. Companies post medical writing job openings on their websites and may also post jobs on the AMWA website.

Large sponsor companies and CROs offer formal training programs on product development topics, experienced senior writers, and an established infrastructure to an entry-level medical writer. Large sponsor companies establish their own standard styles and templates for regulatory documents. In addition, they maintain archives of all completed documents, many of which are stored as electronic, searchable images. Therefore, new medical writers have many resources available as they advance in their careers, under the guidance of more experienced medical writers.

At large CROs, entry-level medical writers also benefit from experienced colleagues. Unlike sponsor companies, though, CROs expect medical writers to follow the clients' document standards. CRO medical writers quickly gain experience with a wide variety of experimental products, therapeutic areas, document templates, formats, and software.

Clients rely on the CRO medical writers to produce professional documents that meet all regulatory agency standards. Sometimes, they also expect the medical writer to advise them on regulatory procedures and requirements. Such client expectations may be challenging for a novice medical writer unless he or she can turn to more experienced colleagues for help.

Mike's best opportunities for medical writing jobs came from networking and personal referrals. More than other clinical specialties, a medical writer's success depends on the quality of the documents that he or she produces. Instructors who are impressed with a student's writing assignments in university writing programs and supervisors who note an employee's written work (such as internal reports) in nonwriting positions can provide influential recommendations when hiring managers are seeking medical writers.

All hiring managers look for the same attributes in candidates for medical writing positions: writing ability, technical knowledge, and interpersonal skills. Mike's writing experience (including published scientific articles, his Ph.D. thesis, and IND preparation) certainly qualifies him for a medical writer position, but there are other ways. Medical writers with a bachelor's degree, master's degree, or healthcare licensure typically start their careers as laboratory technicians, clinical assistants, or healthcare practitioners and drift toward a writing career through opportunity and interest. They demonstrate their writing skills by volunteering for or accepting writing assignments, which often lead to more writing opportunities.

Even for entry-level medical writing positions, clinical hiring managers prefer candidates with clinical experience. By "clinical experience" they mean understanding medical terminology, basic medical concepts, and GCP. Nurses, pharmacists, and other healthcare professionals learn medical terms and concepts during their training. For Mike and others with science degrees in non-medical fields, textbooks, workshops, and online courses offer training on medical terminology, human physiology, pathology, the action of drugs, basic statistics, and grammar. Similarly, GCP training—specifically, the requirements for preparing regulatory documents—is available through industry-sponsored workshops and online training. Hiring managers at medical device companies look for writers who have an understanding of engineering disciplines (such as electronics, software, and biomechanics) that are associated with their products.

BOX 10-3. *Tips for Getting a Medical Writing Position*

- Publish part (or all) of the M.S./Ph.D. dissertation
- Take medical writing courses or workshops
- Learn medical terminology
- Learn basic medical concepts (diseases, human anatomy, pharmacology)
- Take online courses on GCP
- Volunteer for writing tasks (grant proposals, newsletters, abstracts, scientific manuscripts)
- Consider volunteering to write sections of regulatory and clinical documents
- Acquire good keyboarding skills
- Be familiar with current word-processing, spreadsheet, and graphics software (especially document styles and formatting)
- Learn how to create templates
- Assemble a portfolio and include examples of published work
- Join a professional society such as AMWA or DIA
- Network with writing instructors and other medical writers
- Attend professional conferences, such as the AMWA Annual Conference
- Search medical writing job postings on websites of drug, medical device, and CRO companies
- Preferentially, look for entry-level jobs at mid- to large-sized biopharmaceutical companies and CROs

Landing the Job

All of the hiring managers ranked Mike as a highly desirable candidate. Although he had never held a medical writing position, his writing samples impressed them. His articles were logically organized, thoroughly researched, and clearly written.

Mike's writing samples also showed that he had a firm grasp of medical terminology. His answers during his interviews confirmed that he understood medical concepts. In his résumé, Mike listed technical skills that most hiring managers require, such as word processing and experience with software for creating tables, spreadsheets, and graphics.

Finally, hiring managers value medical writers who have good team-work skills, keep cool under pressure, and meet tight deadlines. The documents that medical writers produce require input and review by many team members and other professionals, but the medical writer is ultimately responsible for the document's clarity and timely completion. In discussing Mike's qualifications, his former supervisors and coworkers confirmed that Mike had worked effectively and cooperatively with others on his writing assignments throughout his graduate work and postdoctoral training. He also met the deadlines for his thesis advisor's grant proposals.

The hiring managers were less impressed with the candidates who competed with Mike, even though some included copies of published scientific articles with their résumés. The hiring managers rejected applicants who submitted cover letters and résumés with typographical errors or who were unfamiliar with standard word processing software (e.g., a poorly formatted cover letter or résumé). They also rejected applicants with a history of substandard work, such as not checking references, using inaccurate references, not responding to comments on successive drafts, and not meeting deadlines. Because medical writers must work collaboratively with their clinical colleagues and company executives to decide how to present study results and conclusions, hiring managers also rejected applicants who were inflexible, thin-skinned, or unable to respond constructively to criticism.

Although Mike had never written a clinical study report, he qualified for an entry-level medical writing position at CanDo Pharma. His initial assignments included proofreading and conducting the quality control review of documents written by his coworkers. In the process of performing those tasks, Mike found well-written documents that he could use as models, learned the department's SOPs, and became proficient in using CanDo Pharma's document templates and document management software.

Mike continued his on-the-job training under the guidance of more experienced medical writers and his colleagues in the regulatory affairs department. Having mastered his initial assignments, Mike quickly took more responsibility for writing regulatory documents such as clinical study reports, annual safety reports, investigator's brochures, and sections of MAAs. Although not all good writers are good editors and vice versa, Mike eventually became a skilled editor and helped the clinical authors to polish and improve their draft manuscripts.

To supplement his medical writing experience, Mike joined two professional organizations, the AMWA and Drug Information Association

(DIA). Through its national organization and local chapters, AMWA offers workshops and certificate programs in medical writing. DIA includes an international program track for medical writers at its annual meeting and offers regional workshops on topics specifically of interest to medical writers in the pharmaceutical industry. In addition to attending local DIA and AMWA meetings and workshops, Mike earned a specialty certificate in regulatory and research writing from AMWA.

By the time Mike joined Dr. Abernathy's program on the CanDo Pharma drug, he had written the clinical study reports and other regulatory documents for many other studies. He was fully prepared to work with Dr. Abernathy's strategy team and the various clinical study teams, including Nancy's team, to write the IB and IB updates, protocols, regulatory reports, and clinical study reports.

MOVING FORWARD

Like most medical writers, Mike wanted to advance in his career by continuing to work as a medical writer. Experienced medical writers are in demand, and Mike can consider a variety of medical writing options. Besides regulatory medical writing, he could work as a marketing, medical information, or freelance medical writer.

As a marketing or medical information medical writer, Mike could work at CanDo Pharma, another sponsor company, a CRO, a medical advertising agency, or a healthcare organization. Medical writers in marketing positions write market-oriented documents for consumers and lay audiences as well as professional readers. These include advertising copy, magazine articles, internet content, marketing materials, public relations materials, newsletters, and training manuals.

Writers in a medical information department write product-related medical information for healthcare professionals and consumers. These include standard "medical letters" that summarize available clinical information on the uses of a drug or medical device in specific patient groups, such as the elderly or those with diabetes. Medical information writers also write product monographs and dossiers for payers such as managed-care organizations and insurance companies.

Medical information writers usually hold a degree in pharmacy. Marketing medical writers often have a background in journalism or a degree in English and must acquire scientific and medical knowledge.

Mike's scientific background and clinical experience make him an attractive candidate for these lucrative and interesting positions.

Becoming a freelance medical writer is another option for Mike. Many medical writers, after establishing their credentials by working in one or more corporate positions, turn to freelance writing. Freelance medical writers do the same types of work as regulatory, marketing, and medical information writers. They receive requests from sponsor companies and CROs to write a wide variety of documents and are often given tight deadlines. Although working at home is convenient, freelance writers must maintain office equipment that is compatible with their clients', adapt to their clients' document standards, be proactive in learning up-to-date regulatory requirements, and take considerable initiative to win writing assignments.

Most medical writers continue to develop their writing skills, rather than move into another clinical career track. As an experienced medical writer and with good management skills, Mike may move into a management position in medical writing. Unlike many fields, medical writers who are promoted to senior level or executive positions usually continue to work at their craft, in addition to their responsibilities for leading a medical writing group or department. However, promotion to a management position gives Mike greater freedom to choose his assignments, work on challenging, high-profile documents, and coach less experienced colleagues.

Mike's success and career advancement in all of these areas of medical writing depend on his reputation for writing high-quality documents and, secondarily, his interpersonal skills and ability to meet deadlines. Medical writers who impress one client, whether it is an internal team or an external customer, quickly discover that other clients want to use their services, too. Consequently, medical writers with a good reputation command a steady, lucrative income and can choose their writing assignments.

MEDICAL WRITER RESOURCES

Good Clinical Practices Training and Certificate Programs

ClinfoSource (www.clinfosource.com) offers online GCP courses and certification for CME and CNE credit at reasonable cost.

Kriger Research Center International (www.krigerinternational.com) offers online GCP courses and certification in multiple languages on the country-specific web pages under the "Courses" tab.

Association of Clinical Research Professionals (www.acrpnet.org) offers online GCP courses and certification for CME, CNE, and ACRP credit at reasonable cost.

GCP Training Online (www.gcptraining.org.uk) offers online GCP training and certification, emphasizing the European Clinical Trial Directive, at reasonable cost.

Medical Dictionaries and Terminology

ClinfoSource (www.clinfosource.com) offers an online medical terminology course for CME and CNE credit at reasonable cost.

Kriger Research Center International (www.krigerinternational.com) offers an online course in medical terminology in multiple languages on the country-specific web pages under the "Courses" tab.

Free-ed.net (www.free-ed.net/free-ed/healthcare/medterm-v02.asp) offers an online course in medical terminology and CEU credit at reasonable cost.

Des Moines University (www.dmu.edu/medterms) offers a public access, online course in medical terminology.

Medical Writing Courses and Certificate Programs

American Medical Writers Association (www.amwa.org) offers an extensive catalog of courses for professionals in medical communications and certificate programs in essential skills, composition, science and medicine, and regulatory and research.

University of Chicago (https://grahamschool.uchicago.edu) offers a certificate program in medical writing and editing through the Graham School of General Studies.

University of the Sciences in Philadelphia (www.gradschool.usip.edu/programs) offers an online master's degree in biomedical writing and medical writing certificates through the College of Graduate Studies.

San Diego State University (http://tcomm.sdsu.edu) offers a certificate program in professional writing through the Rhetoric & Writing Studies department.

University of Iowa Rhetoric Project (http://poroi.grad.uiowa.edu) offers a course in medical writing and publishing in the Project on Rhetoric of Inquiry program.

Medical Writer Professional Organizations

American Medical Writers Association (www.amwa.org) promotes excellence in medical communication through educational programs, publications, and networking.

Drug Information Association (www.diahome.org) is an international professional association for those who are involved in discovery, development, regulation, surveillance, or marketing of biopharmaceutical products.

International Society of Medical Publication Professionals (www.ismpp.org) is an international organization for medical publication professionals in the pharmaceutical, biotechnology, and medical device industries.

Profiles of Sponsor and CRO Companies (*See detailed list in Chapter 13*)

Pharmaceutical Research and Manufacturers of America (www.phrma.org) represents the leading pharmaceutical research and biotechnology companies in the United States.

Biotechnology Industry Organization (www.bio.org), the world's largest biotechnology organization, represents more than 1200 companies.

Medical Design & Manufacturing (www.devicelink.com) is an online resource for the medical device industry.

Association of Clinical Research Organizations (www.acrohealth.org) represents the world's leading clinical research organizations.

Salary Surveys for Medical Writers

American Medical Writers Association Salary Survey (www.amwa.org) is published periodically. Access the most recent survey by entering "salary survey" in the website's Search field.

Society for Technical Communication Salary Database (www.stc.org) is published periodically and is available to STC members. Access the most recent survey by entering "salary database" in the website's Search field.

Salary.com (www.salary.com) publishes salary ranges for positions in a wide range of industries. Enter "medical writer" and if appropriate your targeted zip code in the Salary Wizard.

11

Your Future in Clinical Operations

I N CHAPTERS 4–10 YOU LEARNED about the members of a typical clinical study team: their functional responsibilities, their backgrounds, how they landed their entry-level jobs, and their career paths in their current field. This chapter introduces four career paths—clinical study manager, project manager, clinical standards manager, and clinical training manager—that are also available in clinical departments at biopharmaceutical, medical device, and contract research organization (CRO) companies but require previous industry and clinical study experience. These positions, illustrated in Figure 11-1, were mentioned in the earlier chapters as opportunities for career advancement and are described here in more detail.

THE ROLE OF THE CLINICAL STUDY MANAGER

Clinical study managers have overall responsibility for planning, conducting, and completing clinical studies correctly and on time. In some companies, particularly CROs, clinical study managers are also responsible for managing the study's budget. Some clinical study managers supervise a small number of clinical research associates (CRAs) and an administrative support staff in addition to managing clinical studies; other clinical study managers direct the work of one or more clinical study teams without the benefit of reporting relationships. In all cases, the clinical study team members are accountable to the study manager for their work on the study, regardless of their line supervisor.

To manage the study team and its activities effectively, the clinical study manager must have excellent organizational skills and understand

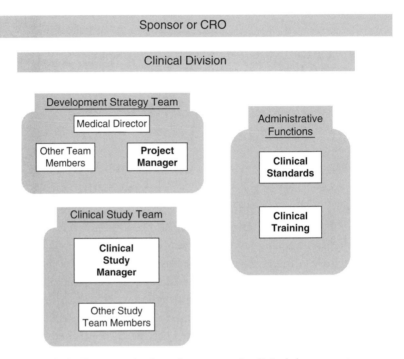

Figure 11-1. Career paths for advancement in clinical departments.

the individual roles of the team members as they plan, conduct, and close a clinical study. In addition, study managers must know how to develop and manage study timelines and have a thorough understanding of global regulatory requirements including Good Clinical Practices (GCPs). Like most clinical study managers, Nancy gained that knowledge and experience by working as a CRA. Others become study managers after gaining experience in project management. Most clinical study manager positions require 7 years of life science or clinical experience, including a minimum of 4 years of industry and product development experience.

Nancy worked closely with Dr. Abernathy to plan her clinical study. With input from Dr. Abernathy's strategy team, she worked closely with Mike, the medical writer, who wrote the clinical protocol and the sample version of the informed consent form. During the internal review and approval process for the protocol, Nancy served as the team's spokesperson, explained and defended the study's rationale, and made adjustments to the study's design and procedures as recommended by the review

---

BOX 11-1. *Clinical Study Manager Responsibilities*

- Plan, oversee, and complete clinical studies
- Manage clinical staff
- Write clinical protocols
- Draft sample informed consent form
- Review monitoring plan, data management plan, statistical analysis plan, and safety management plan
- Facilitate activation of the clinical database and electronic data entry
- Assist in selection of investigators
- Select clinical vendors, CROs, and other study contractors
- Oversee investigator and vendor agreements
- Authorize payments to investigators and vendors
- Manage study budget and resources (especially CRO-based study managers)
- Organize and implement investigator meetings
- Authorize shipment of experimental products to investigator sites
- Lead study team meetings, monitor study progress, resolve problems
- Report study progress for management review and regulatory reporting
- Close out studies
- Assist clinical study report writing
- Manage, train, and evaluate performance of clinical staff

---

committee. She also ensured that the information in the investigator's brochure was current.

Several members of the study team use the information in the protocol to prepare their detailed plans and supporting documents for the study. These include the monitoring plan that Nancy prepared for the CRAs, Maria's data management plan, Tom's statistical analysis plan, and Jasmine's safety management plan. Nancy ensures that those plans are produced on time and according to the corresponding standard operating procedure (SOP) requirements.

Because her study uses electronic case report forms (CRFs), Nancy relies on Maria to complete the programming for data entry and the clinical database before the study starts. She encourages David, the CRA, and the site coordinators to complete their user-acceptance testing of the electronic data-entry procedures quickly, so that Maria can activate the clinical database and electronic data capture (EDC) data collection as soon as possible.

With Dr. Abernathy's input, Nancy compiles the initial list of investigator sites. She works with David, who conducts a site evaluation visit to each site. On the basis of his assessment and the principal investigators' (PIs') responses to CanDo Pharma's survey, Nancy determines whether the sites have the ability to recruit sufficient numbers of patients. For each site that is suitable, qualified, and willing to participate in the study, Nancy works with the CRAs and CanDo Pharma's contract department to negotiate the investigator contract agreements and finalize each site's study budget.

Nancy also selects the vendors and other contractors who are needed for the study. For example, she decides to contract one vendor to assess the X-ray films collected from patients at all the clinical sites in her study. She identifies other vendors to handle other study tasks such as laboratory sample analysis. Nancy facilitates the contract agreements, authorizes payments to the vendors, and oversees their work.

Clinical study managers who work at a CRO also have significant financial responsibility. They are responsible for meeting their client's requirements for completing the study within the contracted budget, as well as on time. If there are unexpected study costs or the client requests additional work, the study manager must negotiate adjustments to the study budget. These budget adjustments are usually not welcomed by the client, and the study manager must be a skilled, sensitive, and persistent negotiator.

With support from Dr. Abernathy, David, Maria, and other team members, Nancy arranges the investigator meeting, including the agenda, meeting location, invitations to participants, confidentiality agreements with participants, and the meeting logistics. By interacting with the PIs and study coordinators at the investigator meeting, Nancy establishes personal relationships that will facilitate discussions during the study, especially when difficult problems must be addressed.

After the investigator meeting and other study preparations have been completed, Nancy—in conjunction with Amy, the regulatory affairs specialist—authorizes shipment of the experimental treatments, both the CanDo Pharma drug and placebo, to the investigator sites. In cases where the sites are initiated individually rather than via an investigator meeting,

Nancy and Amy authorize the drug and placebo shipments after the CRA confirms that the site staff has been trained. Nancy and Amy also comply with regulatory requirements for posting information about the study on websites such as www.clinicaltrials.gov for viewing by patients, healthcare professionals, and the general public.

During the conduct of the study, Nancy makes all operational decisions, referring medical questions to Dr. Abernathy. Often, the PIs encounter situations that the study team did not anticipate and are not addressed in the protocol. In some cases, Nancy simply explains or clarifies the protocol procedures as they apply to the site's specific situation. In other cases, the PI may raise an issue that is more complex and requires a thoughtful discussion with the study team or Dr. Abernathy before giving the PI instructions. Nancy must use her judgment and experience to decide how these issues should be resolved.

Throughout the study, Nancy oversees the study team's progress and leads regularly scheduled team meetings. The meetings are an opportunity for the team to share information and resolve problems.

In addition, Nancy makes periodic progress reports to Dr. Abernathy and others in CanDo Pharma's senior management. She summarizes the status of key study metrics such as the number of patients enrolled and the team's progress in achieving its established milestones, including the projected time when the last enrolled patient will complete the treatment schedule. She also highlights significant problems that may affect the study timelines or study conduct, such as PIs who are having difficulties with finding interested patients, unexpected side effects that emerge during treatment, and logistical problems with the protocol procedures. Nancy communicates how the team is addressing the problems. For solutions that require CanDo Pharma management's approval, she makes recommendations on behalf of the team.

At the completion of the study, Nancy closely follows the work of the CRAs and data management staff as they collect the remaining data and resolve the final CRF queries. She also facilitates important end-of-study activities such as locking the clinical database and reconciling the clinical and safety databases.

During the time that Mike is writing the clinical study report, Nancy is available to answer his questions and coordinate the contributions from other team members. She is one of the key people who reviews and approves the final study report before it is submitted to CanDo Pharma's senior management for their consideration.

Nancy divides her time between multiple study teams and her supervisory responsibilities. Because the workload on studies varies, she may be planning a new study, managing an ongoing study, and reviewing the final study report on another study at the same time. In addition, she assigns work to her staff, provides appropriate training and guidance, and evaluates their performance.

To advance their skills and career networking, clinical study managers usually belong to one or more professional organizations, such as the Association of Clinical Research Professionals (ACRP) and the Drug Information Association (DIA).

THE ROLE OF THE PROJECT MANAGER

Project managers coordinate the activities of the diverse scientific disciplines involved in product development projects and ensure that the projects move forward. Bill serves as the project manager on Dr. Abernathy's strategy team and works closely with Dr. Abernathy to monitor the development program's timelines, milestones, and resources.

In most sponsor companies, the leader of the product development team, like Dr. Abernathy, holds an M.D. degree and is usually a senior-level manager (such as a director or vice president). Development team leaders provide the scientific and business leadership for the entire development and testing program of a new product. Project managers, like Bill, participate in the decision-making process, but the development team leader is usually the decision maker.

At large companies like CanDo Pharma, a project management department takes responsibility for hiring and training project managers and assigning them to product development teams. Bill, an experienced project manager, started working on Dr. Abernathy's team when the strategy team was formed and will continue to manage the program until the product's development is completed. He coordinates the activities of those who manufacture the drug supplies, the laboratory scientists who test the product's safety in animals, and the clinical teams (such as Nancy's) who conduct the clinical studies.

To be effective, Bill must understand the work and interrelationships of the various functional areas, multinational regulatory requirements, how medical products are developed, and CanDo Pharma's business

BOX 11-2. *Project Manager Responsibilities*

- Create, maintain, and document development plans
- Plan and manage program timelines and milestones
- Monitor program resources
- Prepare program status reports
- Track project's progress (such as progress toward milestones and risk mitigation)
- Provide administrative support for product development team
- Schedule team meetings, create agendas, identify action items, record minutes, and follow up ongoing action items
- Facilitate communication to senior managers, team members, and project support staff
- Track expenditures and adjust project budget and staffing requirements (especially CRO-based project managers)
- Show exceptional communication, interpersonal, and diplomatic skills
- Manage people without direct authority
- See the big picture, but also pay attention to the details
- Foster a collaborative, positive work environment
- Show multitasking and organizational skills
- Anticipate difficulties and develop contingency plans

strategy for new product development. He acquired important skills and experience by serving as a member of several clinical study teams. Project managers can also have backgrounds in other areas of product development such as manufacturing, laboratory safety testing, or regulatory affairs.

Most project managers have master's or Ph.D. degrees and 7 years of life science experience, including a minimum of 4 years of industry and product development experience. Some also have credentials from a project management certificate program, such as that offered by the Project Management Institute (PMI) or DIA. However, most biopharmaceutical and medical device companies value project managers with on-the-job product development experience more than project management certification.

The single most important characteristic of successful project managers is communication skills. Often, they must manage people without direct authority, using their ability to persuade, negotiate, and resolve problems. By combining his years of experience in various aspects of product development with exceptionally good interpersonal skills, Bill understands the importance of communicating with all the functional areas and coordinating their cross-functional tasks. Most of all, he ensures that Dr. Abernathy has timely and accurate information, which he needs to lead the strategy team effectively and keep CanDo Pharma's senior management informed.

To monitor the strategy team's progress and plan future activities, Bill relies on a staff of workers in the project management department. Project coordinators (or project planners) assist project managers by entering project plans into a project management database, updating the status of tasks and milestones, preparing standard progress reports, and arranging meetings on behalf of the project manager and team leader. Their assistance with the technical aspects of project management allows Bill to devote most of his time to people-oriented tasks such as sharing project information with Dr. Abernathy and the other strategy team members, following up on action items from team meetings, and resolving conflicts and bottlenecks in the hand-offs from one functional area to another.

Project managers at CROs typically have many years of experience in managing preclinical or clinical studies and have worked in many therapeutic areas for a variety of sponsors. When a sponsor contracts their services, CRO project managers may be responsible for coordinating the work of functional specialists not only at their own CRO and the sponsor company but also specialists at other CROs and independent consultants. Coordination of such complex working relationships requires extensive knowledge of product development activities, outstanding communication skills, and lightly applied persistence. In addition, CRO project managers, unlike their counterparts at sponsor companies, are responsible for managing the project budget and must keep a watchful eye on expenditures and resource utilization.

Because project managers have an excellent understanding of the entire product development process, are familiar with multinational regulatory requirements, and have strong interpersonal skills, they often advance to senior level management positions at sponsor companies and CROs. Many also supplement their training and foster their career advancement by joining professional societies such as DIA or PMI, which offer opportunities for networking and job searches.

THE ROLE OF THE CLINICAL STANDARDS MANAGER

Many large CROs and sponsor companies maintain a standards unit or standards department in their research and development division. Those who work in the standards group manage the company's SOPs. Organizationally, the standards group may reside within a quality department, in which case the group manages all laboratory, manufacturing, and clinical SOPs. Alternatively, the company may have an independent standards function within the clinical department, in which case the group supports clinical SOPs exclusively.

Each clinical SOP covers an activity performed within clinical operations and gives detailed instructions for performing the procedure. Clinical SOP topics include:

- Protocol preparation

- Clinical site selection

- Investigator brochure preparation

- Data collection and recording

- Case report forms

- Study monitoring

- Clinical site auditing

- Adverse event reporting

BOX 11-3. *Clinical Standards Manager Responsibilities*

- Manage life cycle of clinical SOPs
- Provide administrative support to SOP authors and owners
- Store SOPs as controlled documents
- Archive SOPs that have been superseded
- Control SOP access to authorized personnel only

- Statistical review of data

- Clinical study reports

By following the SOP, team members will always perform the procedure the same way. Those who conduct the procedures are responsible for authoring the SOPs in their functional area (for example, clinical data managers write the data collection and case report form SOPs), because they understand the nature of their own work, and they are the ones who must follow the SOP.

Each clinical SOP is written by one or a few authors. However, the SOP topics and corresponding authors cover widely diverse clinical areas, and they benefit from coordination by a centralized standards group. The clinical standards manager serves as a resource for the SOP authors, offering suggestions to ensure that all SOPs follow the same format and include all the elements required by GCPs.

For each new SOP, the clinical standards manager oversees the approval process, ensuring that all SOP stakeholders review, make necessary edits, and approve the SOP. Once approved, the clinical standards manager maintains the SOPs in a secure location, either electronically or as hard copies, and controls SOP access to authorized personnel only.

Regulatory authorities require sponsors and CROs to review and update their SOPs at regular intervals. The clinical standards manager therefore circulates each clinical SOP for review and revision according to its stated review cycle. If the SOP stakeholders recommend revisions, the standards manager coordinates the process of reviewing, editing, and approving the updated SOP. The standards manager officially issues the new SOP, which supersedes the preceding version. The standards group retains all earlier versions of the SOP in the standards archive for future reference, including regulatory inspections.

To be effective in carrying out these responsibilities, clinical standards managers must have a thorough understanding of multinational regulatory requirements, how medical products are developed, and the philosophy of good SOP management. They usually acquire these skills and experience by serving as a clinical quality assurance (CQA) auditor or regulatory affairs specialist and overseeing clinical study teams in a variety of therapeutic areas (see Chapters 7 and 8). Most clinical standards managers have a bachelor's degree or healthcare licensure and 6 years of clinical experience, including a minimum of 4 years of industry and product development experience. Many also supplement their training and foster their career advancement by joining professional societies such as DIA, ACRP, or Society of Quality Assurance (SQA), which offer opportunities for networking and job searches.

THE ROLE OF THE CLINICAL TRAINING MANAGER

Regulatory agencies and International Conference on Harmonization (ICH) guidelines require that each person involved in conducting a clinical study must be qualified to perform his or her respective tasks. As described in Chapters 3–10, clinical workers obtain some of those qualifications through formal education and experience in previous jobs. However, some tasks require workers to have specialized training, which they can best obtain in conjunction with their current job.

Clinical training units at large sponsor companies and CROs conduct training activities to meet those regulatory requirements. Core curriculum courses include training on ICH guidelines, country-specific GCPs, and company-specific clinical SOPs. Because regulatory agencies frequently publish new regulations and companies periodically update their SOPs, the training unit provides ongoing instruction to clinical workers as necessary for their jobs. Sometimes, clinical training managers work in conjunction with managers in the CQA department to conduct the regulatory compliance (see Chapter 7).

Clinical studies often evaluate innovative medical treatments, rely on complicated medical procedures, and address unusual medical conditions.

BOX 11-4. *Clinical Training Manager Responsibilities*

- Develop new courses for training clinical staff
- Prepare course materials
- Coordinate and conduct training sessions
- Track compliance training
- Maintain official training records

To conduct such studies, clinical workers must receive appropriate training on the experimental treatments, medical procedures, and the disease. Clinical training managers, working in conjunction with subject matter experts such as physicians and scientists, create and implement training programs to meet those needs.

Clinical training managers must have a thorough understanding of the product development process, regulatory requirements, clinical SOPs, and at least a basic understanding of the science and technologies underlying clinical studies. Most training managers acquire that knowledge and experience by working as a clinical quality assurance (QA) auditor, study manager, or CRA. Clinical training managers typically have at least a bachelor's degree or healthcare licensure and 7 years of clinical experience, including a minimum of 4 years of industry and product development experience. In addition, they have excellent communications skills and are familiar with adult learners, learning styles, and various instruction techniques including classroom presentations, interactive video instruction, and small group case study exercises.

Regulatory agencies also require sponsor companies and CROs to maintain a clinical training record repository, which archives the training activities completed by each clinical worker. In some companies, the clinical training unit maintains the repository and also maintains lists of training activities that each clinical worker is required to complete for his or her job function. Having a central repository for training records and job-specific requirements makes it easy for sponsor companies and CROs to retrieve and verify the credentials of each clinical worker during regulatory inspections and CQA audits. However, other companies rely on individual departments or the human resources (HR) department to maintain these essential records.

Although all sponsor companies encourage ongoing training of their clinical workers, large CROs often have the most systematic and comprehensive training programs. Because winning sponsor contracts largely depends on having a qualified, experienced, and appropriately trained workforce, CROs invest heavily in training programs and set rigorous skill standards. Consequently, the opportunities for obtaining a clinical training manager position and gaining experience as a trainer are greater at CROs than at sponsor companies. Clinical training managers also supplement their training and foster their career advancement by joining professional societies such as ACRP or DIA, which offer opportunities for networking and job searches.

CLINICAL MANAGER POSITION RESOURCES

Clinical Manager Professional Organizations

Association of Clinical Research Professionals (www.acrpnet.org) is a global resource for clinical research professionals in the pharmaceutical, biotechnology, and medical device industries and those in hospital, academic medical centers, and physician office settings. ACRP provides educational, certification, and networking services for those who support the work of clinical investigations.

Drug Information Association (www.diahome.org) is a worldwide professional association that fosters innovation and exchange of health information. Members include those involved with discovery, development, regulation, and surveillance of biopharmaceutical products.

Project Management Institute (www.pmi.org) is a leading worldwide association that provides educational, certification, professional development, and networking services for project management professionals.

CenterWatch (www.centerwatch.com) is a leading source of news, directories, analysis, educational resources, and proprietary market research for clinical research professionals.

Salary Surveys for Clinical Manager Positions

Applied Clinical Trials Salary Survey (www.appliedclinicaltrialsonline.com) is published annually. Access the most recent survey by entering "salary survey" in the website's Search field.

Association of Clinical Research Professionals Salary Survey (www.acrpnet.org) is published periodically. Access the most recent survey by entering "salary survey" in the website's Search field.

American Association of Pharmaceutical Scientists Salary Survey (www.aaps.org) is published annually. Access the most recent survey by entering "salary survey" in the website's Search field.

Project Management Institute Salary Survey (www.pmi.org) is published periodically. Access the most recent survey by entering "salary survey" in the website's Search field.

Contract Pharma Salary Survey (www.contractpharma.com) is published annually. Access the most recent survey by entering "salary survey" in the website's Search field.

Medical Device & Diagnostic Industry Salary Survey (www.devicelink.com) is published periodically. Access the most recent survey by entering "salary survey" in the website's Search field and selecting the Research and Development survey.

Salary.com (www.salary.com) publishes salary ranges for positions in a wide range of industries. Enter "clinical research manager," "project manager," "quality assurance manager," or "training manager" in the Salary Wizard.

12

Helpful Hints for Landing a Clinical Product Development Job

CHAPTERS 3–10 DISCUSSED THE CLINICAL AREAS in which biopharmaceutical, medical device, and contract research organization (CRO) companies are most likely to recruit entry-level job candidates. For each clinical specialty, you learned about the daily work of a clinical team member, strategies for getting an entry-level job, and the career options that are open to you as you gain experience. In addition, each chapter listed references for professional organizations, training opportunities, salary surveys, and the best places to find jobs in that clinical specialty.

This chapter provides you with practical résumé, application, and interview tips to maximize your chances of getting hired as an entry-level employee in those clinical specialties. The recruiting and hiring process for clinical positions in industry has some unique features. Therefore, this chapter emphasizes these special considerations but also includes some "best practices" that apply to any job search.

START BY GETTING "CLINICAL" INTO YOUR RÉSUMÉ

Based on what you've learned in the preceding chapters, you may have identified the types of clinical work that sound appealing—and also those that you would rather not pursue. So, start your career planning by assessing the gap between your current credentials and those you need to qualify for your preferred entry-level clinical position(s).

Keep in mind that you will always be competing with other eager candidates. Some of them may have previous industry experience, but not necessarily the right type of experience to qualify for a clinical position.

233

Your goal is to do things that allow you to highlight the "clinical" aspects of your résumé. Even if you have no direct experience with clinical studies, hiring managers recognize many other activities as relevant clinical experience for an entry-level industrial job. In the preceding chapters, you learned about the strategies that each CanDo Pharma team member used to gain relevant "clinical experience" that industrial hiring managers accept. Here is a summary of those strategies:

Training

- Basic medical concepts for diseases, human anatomy, or pharmacology
- Medical terminology
- Online courses on Good Clinical Practices (GCP)
- Regulatory requirements posted on International Conference on Harmonization (ICH), International Organization for Standardization (ISO), and Food and Drug Administration (FDA) websites
- Medical recordkeeping
- Medical writing courses or workshops
- Document management system software such as Documentum
- SAS programming certification or Structure Query Language (SQL) training
- Certification as a technician in a medical skill such as phlebotomy
- Certification as a clinical study coordinator

Volunteer

- At a hospital, medical center, or clinic that conducts clinical studies
- At a facility that gives you contact with patients, such as a retirement home
- As a medical record keeper or laboratory scheduler at a medical center or doctor's office
- To assist with data entry or analysis at a medical center
- For writing tasks such as grant proposals and scientific manuscripts
- To assist preparation of regulatory and clinical documents

Relevant Clinical Work Opportunities

- A clinical study coordinator
- An assistant supporting clinical studies at a university or medical center
- An assistant managing patient charts in a doctor's office
- A medical records clerk or office assistant in a hospital or medical center
- A laboratory technician for a physician who conducts independent clinical research
- An institutional review board (IRB) administrative clerk at a hospital or medical center
- A nurse coding medical claims at an insurance company or in a medical practice
- An administrative position at a nonprofit organization, such as Red Cross

"Transferable" Skills Relevant to Clinical Positions

- Conduct an independent study project, analyze the data, and write a report
- Manage and file patient charts
- Read and understand clinical laboratory test results
- Expertise with spreadsheets and relational databases
- Experience organizing scientific data

PREPARING YOUR RÉSUMÉ

Your résumé is how your new employer first learns about you, and it has one purpose: to get you an interview. It is a marketing tool, selling you to the hiring manager as the best person for the job. Make sure the content and presentation of your résumé are easy for the hiring manager to read. If hiring managers cannot find key information, you probably won't get an interview. And without an interview, you won't get hired.

General Tips on Content and Presentation

Many companies use human resources (HR) administrators or employ contract recruiters to screen incoming résumés. These screeners spend less than a minute skimming each résumé and only look for key words and phrases. Hiring managers will look more closely at the résumés that have survived the screeners, but they will not waste time hunting for information. Here are some tips to make your résumé easy for screeners and hiring managers to navigate.

Length and Appearance

- Keep your résumé short, no more than two pages
- Deliver information concisely and avoid repetition
- Use bullets rather than paragraphs of text
- Make only one important point in each bulleted item
- List all dated information in reverse chronological order (most recent item first)
- Use a standard typeface (such as Times New Roman or Arial)
- Use a readable font size (12 point is standard)
- Use italics or boldface type sparingly, if at all
- Leave plenty of "white space" (wide margins and double space between sections)
- Proofread to eliminate all formatting errors
- Do not use fancy borders, clip art, patterns, graphics, multiple colors, extravagant fonts, photos, or fancy formatting with lines, shading, or double columns
- Do not use bound covers or a fancy folder

General Content Considerations

- Include skills and competencies that are relevant to the position; do not give your whole life story
- Display clinical experience prominently. If your experience includes work on clinical studies, list the principal investigator and the type of clinical study, but do not disclose confidential information.

- Highlight your accomplishments. If you are listed as an author on published scientific papers, give the literature citations.

- Back up your achievements with facts that explain their significance. For example, "Increased productivity 2-fold with improved spreadsheet designs" rather than "Designed new spreadsheets."

- Always tell the truth and stick to the facts

- Avoid jargon, abbreviations, and acronyms

- Pay attention to punctuation and use the spell and grammar checkers on your computer

- Always proofread your final résumé to ensure that unusual words and phrases are not corrupted by the computer

If you submit your résumé electronically, create it as a Word or .PDF document and carefully follow the company's submission instructions to ensure that the document is not corrupted when it is received and opened.

If you submit your résumé as a hard copy, print an original from your printer; do not send a photocopy. Use high-quality, white paper and secure the pages with a single staple in the corner. Internally, the company will probably make multiple photocopies of your résumé, which should remain legible when duplicated on a monochrome copier.

Standard Résumé Headings: Some General Suggestions

Hiring managers search for two things in your résumé: specific technical skills that match the position's requirements and a work history of achievements. They are accustomed to seeing certain headings in a certain order. Below is a standard list of résumé headings that make it easy for the hiring manager to navigate. However, use only headings and subheadings that apply to you. If necessary, modify the headings to ensure that they clearly and accurately reflect your skills and accomplishments.

- Contact information: At the top of the first page, list your name, mailing address, home and cell phone numbers (day and evening), and email address. Include your social networking address, but only if your webpage displays your "professional" side.

- Summary: In no more than two sentences, give the reader a succinct assessment of your strengths as a candidate and what you hope to achieve on the job.

- Education: Starting with your most recent (or present) formal education, list the institution name, degree awarded, major (and minor) field, and date awarded. If you received any science prizes, awards, or scholarships, list them.

- Training: List relevant, nondegree training courses, including any resulting qualifications and earned credits, such as continuing medical education credit.

- Professional qualifications: List any professional certifications or licenses and memberships in professional societies (even if you are a student member).

- Employment: Start with your most recent (or current) employer. Summarize your major achievements in each job and any specific benefits you have given to the company. (If you have not yet worked full time for an employer, eliminate this heading and summarize your work experiences under the Experience heading.)

- Experience: Starting with your most recent (or current) activities, list part-time jobs and unpaid work experience. List only those that are relevant to the job you are applying for. For each work situation, include your main responsibilities, transferable skills that you have acquired, and your major achievements. Be sure to quantify your achievements with facts and figures.

- Specific skills: The content and size of this section will depend on the type of work you have done and the type of job you are seeking. Include computer skills (including software packages), clinical skills (e.g., phlebotomy), and relevant on-the-job training.

- Additional information: This section is for you to add information on other achievements, such as language skills, international experience, management experience, and other activities that enhance your attractiveness as a candidate and sharpen the hiring manager's impression of you. If you have done voluntary or charitable work (such as fundraising, a leadership role, or data entry), add it here. Emphasize only those qualities that apply to the type of position you are seeking. For example, being captain or manager of your school soccer team is relevant if you are working toward a leadership role.

- Hobbies: Include only things that support your qualifications. Properly selected interests offer the hiring manager some insight on how well you'll perform in the position, especially if you lack direct professional experience. (Reading is great but doesn't show your teamwork skills; participation in team sports does.)

- References available on request statement (see References section, below).

Things to Avoid in Your Résumé

Avoiding pitfalls is just as important as the information you include in your résumé. Here are some suggestions when preparing your résumé:

- Do not attempt humor. Your résumé is a serious, business document, and you want the hiring manager to take you seriously.

- Do not list failures of any kind: businesses, exams, or personal failures.

- Do not include any personal details: age, marital status, health conditions, or photos. Equal employment opportunity legislation prevents employers from using personal information in their selection process, unless you voluntarily share it.

- Do not include your salary expectations. This is an area of negotiation, and your bargaining position becomes stronger after you begin interviewing (see below).

- Do not exaggerate your accomplishments. You may be asked to back up your claims with specific facts or examples; make sure you can.

- Do not sell yourself short. Be honest but do not be shy. Your résumé is your only opportunity to convince the HR recruiter and hiring manager that you are the right person to invite for an interview.

Templates you can use to construct your résumé and examples of properly formatted résumés can be found on the websites listed in the Resources section below.

FINDING, APPLYING, AND FOLLOWING UP
ON JOB POSTINGS

Finding Clinical Jobs: Getting Started

The easiest and fastest way to gain clinical study experience is in a nonindustrial clinical setting. Investigators who conduct clinical studies or manage clinical research laboratories at major medical centers are always looking for support staff. To find clinical study investigators, browse the government website, www.ClinicalTrials.gov, or ask your personal physician about ongoing clinical studies in your area. Also, contact the human resources office at large hospitals and medical centers to learn about the clinical opportunities available at their institutions.

The application, interview, and negotiating process for nonindustrial positions is faster, less cumbersome, and usually less formal than for industrial positions. Often, just being at the right place at the right time and asking the right questions will link you to a rewarding clinical job opportunity. To maximize the likelihood of being in the right place at the right time, consider offering to work as a volunteer in a physician's office, medical center, or hospital. You will gain valuable skills, and you can use your performance and insider's connections to impress the hiring manager when a job becomes available.

Large CROs are also a good way to begin your clinical career. The corporate headquarters office of each CRO offers the greatest variety and number of entry-level clinical opportunities. (See Chapter 13 for a list of CROs.) For entry-level positions, CROs require little or no previous clinical experience and offer excellent clinical training programs. However, their staffing needs vary, based on the number of contracts they have signed with biopharmaceutical and medical device clients. When they sign a contract for new work, CROs will hire additional staff quickly. However, the application, interview, and negotiating processes are similar to those for other industrial positions (see below).

Obtaining work through a temporary workforce agency is another way of starting your clinical career. Some agencies (such as ACT-1, KFORCE, and i3 Pharma Resourcing) specialize in placing temporary workers in clinical positions at biopharmaceutical, medical device, and CRO companies. (See Chapter 13 for listings of temporary workforce agencies.) If you have already developed some clinical skills through nonindustrial jobs or online training, the agency is more likely to give you a clinically relevant assignment. Do impressive work, and your temporary boss may become your full-time boss or may recommend you to a colleague.

Professions within the medical products industry have their own associations, and most of these have journals, websites, and local chapters. (See Chapter 13 for a list of professional and trade organizations.) Recruiters advertise in the associations' journals and on websites to target specific types of job candidates. If you have narrowed your clinical interest to a specific clinical area (such as data management or regulatory affairs), contact the appropriate association and see if you can become an affiliate or student member. Take advantage of the association's training and career development programs and if it has a local chapter, attend the meetings regularly. Networking is a good way to find out about new jobs.

Use your university's alumni network or a professional online network such as LinkedIn to identify employees who currently work at the companies you are interested in. Make an appointment to speak to these employees and to learn about their job and their company. Your goal in these conversations is not to ask for a job but rather to establish personal contacts with influential people who can advise you about your job search and refine your job search strategy.

Finally, the websites of industry trade organizations, recruiting firms that specialize in medical product development positions, and individual biopharmaceutical, medical device, and CRO companies all include career opportunities home pages and post clinical job openings. (See Chapter 13 for the website addresses.) You can search these online job listings by location, job title, salary range, and other criteria.

While you are gaining clinical skills and searching for job openings, grab every opportunity to become acquainted with people who are experienced in clinical research. They can facilitate your contacts with hiring managers, inform you about job openings before they are posted, and provide influential recommendations when the hiring manager checks your references.

Expect to spend many months finding and landing a clinical position in industry. For hiring managers, the logistics of systematically evaluating résumés, scheduling interviews, gaining approval to make an offer, and negotiating with the top candidate all take time. Adding to this complex process, hiring managers are constantly distracted by internal responsibilities that further delay their progress in filling new positions. Unfortunately, as companies change their priorities, many job postings

appear and then quickly disappear without being filled—further complicating your progress in being hired. Persistence is the key to finding a job.

Applying

Feel free to apply for positions you are interested in, even if you do not
have all of the qualifications. As long as you can include some clinically
relevant experience, training, or skills in your résumé, you have a chance.
You won't get a position if you do not apply.

However, take the time to tailor your résumé for each specific job
application and employer. Read the job description and requirements
carefully and adjust your résumé to highlight achievements that clearly
show you match those requirements and are a good fit for the job. Include
key words and phrases that are used in the job posting; many companies
use electronic filters that search for matches with key words in their job
descriptions. For entry-level positions, hiring managers look for candidates who have "transferable" skills. Be sure to highlight those areas in
your background—especially clinically related activities—that show you
have the interest and potential for doing the job successfully, even though
you have not done it before.

It may seem like a lot of work, but if you emphasize clinical skills that
relate specifically to the job's requirements, you are more likely to get an
interview. Always keep a copy of each customized résumé so that you
know which version you sent to each employer.

Before you submit your résumé, do some additional homework.
Learn about the company, department, and if possible, the hiring manager. Your résumé is more likely to be noticed and considered if you send
it directly to the hiring manager or the department head. To identify
them, use the contacts you have established via your university alumni
network, your professional association memberships, clinical training
courses, or while working at a clinical study site.

If possible, arrange for an employee referral. Many companies offer
their employees cash incentives for finding candidates who are subsequently hired. If you send your résumé to the hiring manager or department
head, reference the employee who referred you. Your résumé is more likely to get noticed, and your employee contact may be financially rewarded.

If you send your résumé as a hard copy, follow the guidelines above
for formatting and printing your résumé and include a cover letter.

Because the cover letter of your hard copy résumé is the first thing the recipient sees, make it a strong sales document. Include the job requisition number and a brief description of the position you are seeking. If someone recommended that you send your résumé to this particular recipient, mention the person by name. Highlight your key clinical skills and show how well you match the requirements for the posted job, but keep the discussion brief. Conclude by noting that you will call to check on the progress of your application. Your cover letter should be one page and use a standard letter format, the same type of paper, and the same font as your résumé.

If you e-mail your résumé to your employee contact at the company or a specific manager, your e-mail message is the equivalent of a cover letter; include the same cover letter information (i.e., the job requisition number, highlights of your qualifications, and justify their relevance to the job). "Attached please find my résumé" is not sufficient.

Keep the format of your electronic résumé simple and use either MS Word or a .PDF image. Display your name and contact information prominently on the first page of the document, not in a header, footer, or special text box. Save your document with a name that includes your name and the current date so that the hiring manager and human resources staff will be able to store and find your résumé easily. Include your résumé as an e-mail attachment; do not embed it in the text of your e-mail. Finally, test your formatting by sending your documents to your, or a friend's, Blackberry; many recruiters access incoming job applications remotely, and some handheld devices have limited formatting options.

If you decide to apply via an online website rather than by hard copy or e-mail, go directly to the employer's website. The job information on the websites of biopharmaceutical, medical device, and CRO companies is usually more current and accurate than that on the websites of recruiters and professional associations. However, do not substitute technology for quality and focus, just because it's easy. The internet is certainly speedy and convenient, but your résumé can easily get buried among thousands of other general résumés, and a recruiter or hiring manager may never find it.

Make your website application specific, targeted, and appropriately formatted, just as you should for mailed résumés. When you submit your résumé via an employer's website, target a specific position, including the job requisition number. Some electronic applications require you to enter the details of your résumé by typing information into the fields of a standard electronic form. However, if the website permits you to attach your résumé as an electronic document, strictly follow all rules specified on the website for electronic submission, including formatting requirements.

Most employers allow you to apply for multiple jobs, but be selective. A company may have multiple positions available with the same job title, each with a unique requisition number. Applying to each of those job requisitions increases your chances, because several hiring managers have posted similar or identical positions. However, limit the number of applications you submit to positions that have different job titles and descriptions. Hiring managers can search their jobs database and see all of your submissions. If you apply for everything, they will probably disregard your application. They want to hire someone who is specifically interested in their job opening, not someone who is fishing.

Finally, keep a list of where and when you sent your résumé, along with copies of the customized versions you used.

Follow-up

If you have sent your résumé to a specific person and indicated in your cover letter (or e-mail) a timeframe for following up, make sure that you do. This is an opportunity for you to confirm that your résumé was received, determine the status of your application, and obtain information about the recruiting process. Be diligent, but do not annoy the person with too many questions or repeated phone calls.

Periodically check the company's website to see if the job requisition number is still listed. If not, the job has either been filled or withdrawn. If the job is still posted and you remain interested in the company but your résumé and contact information have changed, you may wish to submit an updated résumé. Your new résumé will be associated with a new application date. Remember that companies now track the history of all job applications in their recruiting database, and a hiring manager can easily spot applicants who have applied multiple times.

INTERVIEWING FOR A CLINICAL POSITION

Your résumé will be reviewed by a contract recruiter, the company's HR manager, or the hiring manager for just one thing: to decide whether to invite you for an interview. If your résumé highlights the clinically relevant details of your skills, training, and experience in a way that matches the position's requirements, you stand a good chance of being invited for an interview.

Industry employers invest a lot of time and effort in interviewing job candidates. To narrow their focus to only the best candidates, most hiring companies conduct a preliminary telephone interview to screen applicants before making the decision to schedule a face-to-face interview.

Preparing before the Interview

Preparation is important, whether your discussions with representatives of the hiring company are via a telephone screen or a face-to-face interview. The decision to hire you versus other candidates is based mostly on the impression you make during those discussions. Here are some tips to guide your interview preparation:

- Research the company. Review the company's latest annual report, press releases, and scientific announcements. The people you meet will be justifiably proud of their company and its accomplishments. You will impress your interviewers if you show that you know the company and its reputation.

- Review the job description and your résumé. For each skill and accomplishment that you highlighted in your résumé, be prepared to give details of your experience. For entry-level jobs, hiring managers look closely at "transferable" skills; that is, skills that you may have learned outside the workplace or in a different setting but can apply to the recruited position. Be prepared to describe skills that you can transfer to this new job. Your interviewers will be especially interested in your clinical activities. Be prepared to describe exactly what you have done in a clinical setting and what you learned.

- Have a salary range in mind. Consult salary surveys (see Chapter 13) for this clinical job category and gather input from the professionals you have met through your alumni network, during your training courses, and at professional association meetings.

- Ask for an agenda. For a telephone screen, ask about the discussion format and identify to whom you will be speaking. For face-to-face interviews, try to obtain the interview schedule and determine who the interviewers will be by name, title, and department. Be sure to bring the agenda along to your interview.

- Prepare a list of questions for each interviewer. Asking thoughtful questions that are relevant for individual interviewers shows that you did your homework and are interested in the job.

- Have a copy of your customized résumé handy. It will remind you of the details you included in the version your interviewers are using. For your face-to-face interview, be prepared to distribute copies of your customized résumé if requested.

- Bring along business cards. If you have a job, you may have a job-related business card. If not, consider having personal "business" cards printed, showing your contact information.

- Bring examples of your work, if relevant. Work samples are tangible proof of your accomplishments, but make sure it is your work and exclude anything that is confidential.

- Organize your papers for easy retrieval. Give the interviewers the impression that you have everything at your fingertips. For face-to-face interviews, carry your résumé, business cards, interview agenda, and work examples in a brief case or a business organizer.

- Obtain directions to the interview location. For telephone interviews, determine who will initiate the call. If you are driving to the interview, make sure you have clear directions. Know where to park and where to check in. Make sure you arrive on time, and contact your interviewer if you are unavoidably delayed.

Telephone Interviews

For telephone screens and interviews, find a quiet place to take the call, where you will not be interrupted. Use a land line rather than a wireless phone, to ensure that you maintain a good, clear connection throughout the conversation. If your telephone service has a call waiting feature or multiple lines, avoid the temptation to interrupt the interviewer by taking another call.

Telephone screens are usually conducted by one person: a contracted recruiter, HR manager, or sometimes the hiring manager. Assume that your interviewer has carefully read your résumé. He or she will likely ask you to explain or elaborate on selected items in your résumé, so be prepared to answer the questions directly and with specific examples. Be honest and acknowledge the limits of your responsibilities and experiences. An interviewer can easily spot applicants who make claims about things they actually did not do.

Because of scheduling conflicts on the day that you go to your face-to-face interview, you may be asked to speak to one or more interviewers by telephone. If this happens, try your best to accommodate these interviewers. Clearly, they are very busy people, but despite their heavy work schedule someone thinks it is important for them to interview you. They probably will be influential in making the final decision about whom to hire, but unfortunately they will only be able to judge your abilities by what they hear over the phone. Follow the guidelines for a face-to-face interview (see below), but be especially mindful that your voice must do all the work of convincing these interviewers of your fitness for the job.

Face-to-Face Interviews

Interviews for positions at biopharmaceutical, medical device, and CRO companies are grueling, consisting of a full day of meetings with six to ten interviewers. Candidates who pass the first day of interviews often must face a second round of interviews a few days or weeks later. Your interviewers will typically include those who will become your new coworkers and department managers, in addition to the hiring manager and the HR manager. If the position requires interdisciplinary teamwork, you may also meet with representatives of the related departments. Most interviewers will offer you their business cards; if they do not, ask for one. The contact information will be useful for future reference. Before you move to your next appointment, be sure to leave your business card and thank the interviewer for his or her time.

Most clinical interviewers assume that you have the qualifications for the job, because they only schedule interviews with candidates who have strong résumés and pass the phone screen. They see little point in discussing your résumé further. Their purpose in seeing you is to assess whether you will "fit in" with their organization, and they will ask questions to assess that rapport. Be honest, be yourself, but foremost, be businesslike and professional. Take your cues from your interviewers. If they are informal and make jokes, you can loosen up; if they do not, just be respectful and courteous—even if it means sitting quietly and spending most of the time listening.

Many clinical interviewers like to talk. Let them, but try to find opportunities to promote yourself. Make your remarks concise and relevant to the topic your interviewer is discussing. If you agree, say so. If you disagree, be tactful and be prepared to give specific information to support your opinion, but do not argue. Most importantly, give examples of your own work or experience that is relevant to the topic. The interviewer may not want to pursue it, but at least you have communicated something about yourself.

When you speak, maintain good eye contact and be enthusiastic. Avoid saying anything negative, especially about your current or former supervisors. If the conversation drifts to an unpleasant previous experience, be prepared with an honest, brief, and upbeat response.

Some of your interviewers may pepper you with questions. Be prepared to sell yourself by giving convincing answers that prove you are the best person for the job.

Practice your answers to the common questions they might ask:

- What are your strengths and weaknesses?

- How do you manage your time for efficiency?

- Describe how you deal with a "difficult person."

- Why should we hire you for this job?

- Why do you want this job?

Your interviewers may ask you questions about the things you have highlighted in your résumé. For those questions, be prepared to give a specific example that illustrates your skills or work, and try to choose examples that have a happy ending or positive outcome. Also, be prepared to give specific examples in response to general questions, such as:

- What has been your greatest challenge at work?

- What is your greatest achievement?

- What was the hardest thing you've been asked to do? Why was it hard?

- Describe your contributions on a team (or in leading a team).

- What kinds of problems do people ask you to solve?

Most interviewers will give you an opportunity to ask questions. Have your prepared questions handy but also jot down questions that occur to you during the interview. Your questions will expose your level of interest and your personal priorities, so be thoughtful in what you ask and how you ask it.

Potential questions might be:

• In this position, what is a typical day like?

• How do you measure success?

• What do you see as the biggest challenges for people in this position?

• What are your department's (or team's) near-term goals?

• What role will the person in this position have in achieving your near-term goals?

As an alternative to individual interviews, you may meet with a group of interviewers. These panel interviews are popular, because they offer greater scheduling efficiency. However, panel interviews are fast paced, and you will have less time to reflect between questions. Also, the panel will probably have a set of questions that they ask each candidate, and your answers will be compared with the other candidates'. Do your homework, focus your answers, relax, and most of all make sure your answers shine.

One of your interviewers will be an HR representative, who will review the company's benefit programs and other personnel policies. Feel free to ask frank questions about compensation, healthcare, retirement, and savings plans. Most companies in the healthcare industry offer attractive compensation and benefits, but you should understand them.

If the HR representative asks about your salary expectations, provide a salary range that is acceptable to you. Many companies try to gather information about your salary requirements early in the recruiting process, to ensure that you and the hiring manager are thinking in the same salary range; it avoids disappointment and wasting everybody's time. However, do not give an exact figure during your interviews. You will have an opportunity later to negotiate the most favorable terms for your salary and benefits package. To assist you in those negotiations, ask the HR representative about the company's philosophy regarding salaries, bonuses, and promotions.

Before you leave, ask the status of the interviewing process (i.e., are other candidates scheduled to interview?), the next steps for considering you as a candidate, and the timeframe for filling the position. Unless you have decided you definitely do not want to work at this company, show your enthusiasm and express your interest in the position.

In many cases, the next step in filling clinical positions is a second round of interviews. You may meet a second time with the hiring manager and HR representative, but the other interviewers are usually more important decision makers than the interviewers you met during the first round. You should approach these interviews the same way and make certain that you give the same answers, even if you are asked the same questions multiple times. Interviewers compare notes, and if you give conflicting answers, you weaken your chances of being hired.

After the Interview

After the interview, follow up with short thank-you notes, either hand-written or e-mail, to your interviewers—at least the key people you met. Be courteous, thank them for their time, and indicate your level of interest in the position. If you are not at all interested after seeing the company and meeting the people, be honest and respectfully withdraw yourself from further consideration. Otherwise, be positive and give specific examples of the things you saw or heard during your interview that reinforced your interest in the position.

Clinical positions are filled only after a long and careful selection process. Internal activities often shift priorities, and the logistics of scheduling candidates for interviews may extend the recruiting timetable. It is appropriate for you to follow up, but only if you have not been contacted within the timeframe they indicated during your interview.

JOB OFFERS: WHAT TO EXPECT

Before making a job offer, hiring managers and HR managers do their homework. If they are seriously considering you for the job, they will ask you to provide a few references. Choose people carefully and focus on those who can confirm and explain the clinical skills and "transferable" skills you have acquired. Include at least one of your supervisors and one previous or current coworker, unless you are asked to submit the names of other types of people.

The hiring company's homework includes conducting a background check of your credit, criminal, and academic history. Clinical employers are careful to confirm that you have been responsible in paying your bills, complying with laws, and representing your education honestly. Most companies also will require that you successfully pass a screening test for illegal drugs before finalizing your employment.

While they are completing their homework, the HR manager or recruiter will probably contact you to discuss the job offer. This is an important negotiating session. The goal is to discuss compensation and benefits terms that are agreeable to you and your new employer. The offer letter, when it comes, should reflect the verbal agreement from this discussion and contain no surprises.

Keep in mind that salaries are determined by a formula that must meet standards for fairness and competitiveness across all job categories at the company. You will probably have difficulty pushing for additional salary beyond a certain limit. You can get a good idea of what that limit should be by consulting salary surveys for that type of clinical job. However, you can easily bargain for additional benefits, such as a sign-on bonus, relocation expenses, extra vacation time, and perhaps stock options. Be reasonable, be willing to compromise, and be prepared to justify your requests, but do not miss this important opportunity. Remember, they called you to ask what it would take to hire you; now is not the time to be shy.

At the close of the telephone discussion (and it may actually be several conversations), you and the hiring manager should have reached one of two conclusions. If your negotiations were successful, the company will issue a written job offer, knowing that the terms will be acceptable to you. Alternatively, if you do not like the terms, cannot get the company's representative to agree to your requirements, or simply do not want the position, courteously say so and withdraw from consideration at this point.

Of course, when you receive the written offer letter, you are still free to reject it. But unless the written offer contains terms that you were not expecting and are unacceptable, it is considered bad form to decline a negotiated offer. Based on the verbal agreement, the hiring manager and his or her department assume you will be joining the company. Even before they get your written acceptance, they will begin making plans for you to start work.

THE SMALL STUFF

Clinical departments at biopharmaceutical, medical device, and CRO companies must comply with many rules and regulations. Be aware that hiring managers are trained to notice small things that you might otherwise overlook or not consider important. These small things are important, and during the interviewing and recruiting process they contribute to the hiring manager's decision whether to hire you.

Interview Etiquette and Dress Code

Most companies have casual dress codes and most of your interviewers will be comfortably dressed. Nevertheless, you should wear standard business attire, unless the person who invited you gives you other instructions. Business clothes show that you want to make a good impression.

Men should wear a suit (or sport coat) and tie, and women should wear a suit, daytime dress, or pant suit. Assume that you will be doing a lot of walking, so make sure your dress shoes are comfortable. Wear modest jewelry; body piercing, other than earrings, is usually frowned on. Keep fragrances to a minimum. If you have tattoos, wear clothing that covers them.

Arrive early and use the time before your interviews to settle your thoughts, read through your notes, and scan company literature if it is displayed in the waiting room. Turn off your cell phone and other electronic devices and avoid the temptation to make or answer any calls or text messages during your interviews. While you are meeting with your interviewers, show them that they are more important than anyone else.

Be friendly to the receptionist, secretaries, and other administrative staff. They are often asked for their opinion about candidates and have lots of influence in either facilitating or blocking your chances of getting the job.

References

Only include references in your résumé if the job posting specifically requires them. Instead, indicate "references available on request" at the end of your résumé.

Maintain a list of people who can serve as potential references. The list should include previous supervisors, coworkers, and if you led a team or group, key members. Include only people who are very familiar with your work, especially your clinical duties, and will recommend you enthusiastically. Ask their permission and use them as references only if they agree to speak on your behalf.

Hiring managers have different views about the types of people making recommendations and the kind of information they want to know about you. When they ask you to provide references, always alert your references to let them know you will be submitting their names. If possible, tell them who will be contacting them, the approximate timeframe, and whether the communication will be by letter, e-mail, or a telephone conversation. Also, follow up with your references to let them know the status of your interviews. Sometimes, the hiring manager does not contact all of the references you provide, but you should thank each reference, whether the hiring manager contacted them or not.

Personal Websites and Social Networks

If you have a page on a social network such as MySpace or Facebook, be aware that recruiters and HR managers are likely to review it as part of their pre-offer homework, whether you list the address on your résumé or not. Remove all photos, personal details, and other items that do not present you as a serious candidate for the position.

Alternatively, take advantage of a professional network such as LinkedIn, which allows you to post your professional profile and references. Direct hiring managers and recruiters to this site, and they will be less likely to surf for the personal networks that might hurt your chances for serious consideration. Recruiters also use LinkedIn, in addition to professional society membership directories, as a resource to find possible job candidates.

E-mail versus Texting versus Formal Letters

Be professional in all of your communications with recruiters, HR managers, and hiring managers. Whether your communications are in hard-copy or electronic format, avoid using "text language." Remember the rules of grammar: start your sentences with a capital letter, use complete sentences, avoid abbreviations, and end with a period. Correct all spelling and grammar.

RESOURCES FOR PREPARING A STRONG CLINICAL RÉSUMÉ

You can find tips, templates, and examples of properly constructed résumés on the following websites:

Jobs in Science: CV Writing (www.jobsinscience.com/info/cv.asp)

Quality Professional Resume and CV Samples (www.resumesandcoverletters.com/sample_resumes.html)

Job Search: Writing Curriculum Vitae (www.jobsearch.about.com/cs/curriculumvitae/a/curriculumvitae.htm)

OWL at Purdue: Writing the Curriculum Vitae (http://owl.english.purdue.edu/owl/resource/641/01)

JobStar Resume Guide—Curriculum Vitae (www.jobstar.org/tools/resume/res-cv.php)

13

Reference Materials

BIOPHARMACEUTICAL, MEDICAL DEVICE, and contract research organization (CRO) companies post information about job openings on their company websites (see Tables 13-1, 13-2, 13-3, and 13-4). Look for a link or tab labeled "Careers," "Job Seekers," or similar description. This link or tab often launches a careers page that displays general information about the company's corporate goals and work environment. For large companies that have many divisions, the careers page may also display a pick list that will allow you to refine your search by selecting individual corporate divisions. Companies use a variety of names to describe their clinical research and development (R&D) function. Look for a category called "Clinical Development," "Late Stage Development," "Development," or "Research & Development."

On the careers page or business division page, you will see a link or tab that allows you to conduct an online search for open positions. It may be called "Job Search," "Search Openings," or "Search Jobs at (company name)." The search engines vary, but most will allow you to select criteria to narrow your search by geographic area, department, job title, and key words. Clinical positions are typically listed under a category such as Clinical Development, Clinical Affairs, Clinical Operations, Research & Development, Late Stage Development, or Development. In some companies, the pick list may include specialized clinical positions such as regulatory affairs and quality assurance among the categories. Because job titles vary considerably between companies, consider not restricting your search to a specific job title, but rather search on all available clinical positions using the geographic, business division, and department categories. Narrow the search to your area of interest (e.g., data management, contract research associate [CRA], or clinical safety) using the key words field of the search engine.

Table 13-1. *Top Pharmaceutical Companies*

Abbott Laboratories	www.abbott.com
Alcon Inc.	www.alcon.com
Allergan Inc.	www.allergan.com
Astellas Pharma Inc.	www.astellas.com
AstraZeneca Plc.	www.astrazeneca.com
Baxter International Inc.	www.baxter.com
Bayer AG	www.bayer.com
Boehringer Ingelheim GmbH	www.boehringer-ingelheim.com
Bristol-Myers Squibb Co.	www.bms.com
Chugai Pharma	www.chugai-pharm.com
Daiichi Sankyo Co.	www.daiichisankyo.com
Dianippon Sumitomo Pharma Co.	www.dsp-america.com
Eisai Co.	www.eisai.com
Eli Lilly and Co.	www.lilly.com
Forest Laboratories Inc.	www.frx.com
GlaxoSmithKline Plc.	www.gsk.com
H. Lundbeck AS	www.lundbeck.com
Hospira Inc.	www.hospira.com
Johnson & Johnson	www.jnj.com
Kyowa Hakko Kirin Pharma	www.kyowa-kirin-pharma.com
Merck & Co.	www.merck.com
Mitsubishi Tanabe Pharma Corp.	www.mt-pharma-america.com
Novartis	www.novartis.com
Novo Nordisk AS	www.novonordisk.com
Nycomed Group	www.nycomed.com
Otsuka Pharmacuetical Co.	www.otsuka-global.com
Pfizer Inc.	www.pfizer.com
Roche	www.roche.com
Sanofi-Aventis Group	www.sanofi-aventis.com
Shire Plc.	www.shire.com
Solvay SA	www.solvay.com
Takeda Pharmaceutical Co.	www.takeda.com
Watson Pharmaceuticals Inc.	www.watsonpharm.com

Another way to start your clinical career is to register with a temporary workforce agency that places workers in clinical positions at drug, medical device, and CRO companies (see Table 13-5). Clinical managers often contact these firms to fill immediate, short-term work needs, rather than hiring full-time employees. Temporary workforce agencies primarily receive requests for entry-level and junior-level clinical workers, and therefore this is an effective way to launch your clinical career.

Table 13-2. *Top Biotechnology Companies*

Abraxis BioScience Inc.	www.abraxisbio.com
Actelion Ltd.	www.actelion.com
AEterna Zentaris	www.zentaris.com
Alexion Pharmaceuticals Inc.	www.alexionpharm.com
Alkermes Inc.	www.alkermes.com
Amgen Inc.	www.amgen.com
Amylin Pharmaceuticals Inc.	www.amylin.com
Array BioPharma Inc.	www.arraybiopharma.com
BioCryst Pharmaceuticals Inc.	www.biocryst.com
Biogen Idec Inc.	www.biogenidec.com
BioMarin Pharmaceutical Inc.	www.biomarinpharm.com
BioSante Pharmaceuticals	www.biosantepharma.com
Celgene Corp.	www.celgene.com
Cephalon Inc.	www.cephalon.com
Crucell NV	www.crucell.com
CSL Ltd.	www.csl.com.au
Cubist Pharmaceuticals Inc.	www.cubist.com
Elan Corp.	www.elan.com
Enzon Pharmaceuticals Inc.	www.enzon.com
Exelixis Inc.	www.exelixis.com
Facet Biotech	www.facetbiotech.com
Genmab AS	www.genmab.com
Genentech Inc.	www.gene.com
Genzyme Corp.	www.genzyme.com
Gilead Sciences Inc.	www.gilead.com
Human Genome Sciences Inc.	www.hgsi.com
ImmunoGen	www.immunogen.com
Intercell AG	www.intercell.com
InterMune Inc.	www.intermune.com
Isis Pharmaceuticals Inc.	www.isispharm.com
Lexicon Pharmaceuticals Inc.	www.lexpharma.com
Myriad Genetics Inc.	www.myriad.com
Nektar Therapeutics	www.nektar.com
NeuroSearch AS	www.neurosearch.com
Onyx Pharmaceuticals	www.onyx-pharm.com
OSI Pharmaceuticals Inc.	www.osip.com
Progenics Pharmaceuticals Inc.	www.progenics.com
Regeneron Pharmaceuticals Inc.	www.regeneron.com
Seattle Genetics Inc.	www.seagen.com
UCB SA	www.ucb-group.com
United Therapeutics Corp.	www.unither.com
Vertex Pharmaceuticals Inc.	www.vrtx.com
ViroPharma	www.viropharma.com
Xoma Ltd.	www.xoma.com
ZymoGenetics Inc.	www.zymogenetics.com

Table 13-3. *Top Medical Device Companies*

Abbott Laboratories	www.abbott.com
Alcon	www.alcon.com
Baxter International	www.baxter.com
Beckman Coulter	www.beckmancoulter.com
Becton, Dickinson & Co.	www.bd.com
Boston Scientific	www.bostonscientific.com
Cardinal Health	www.cardinal.com
Covidien	www.covidien.com
C. R. Bard	www.crbard.com
Fresenius Medical Care	www.fresenius.com
GE Healthcare	www.gehealthcare.com
Hospira	www.hospira.com
Johnson & Johnson	www.jnj.com
Siemens Healthcare	www.medical.siemens.com
Medtronic	www.medtronic.com
Philips Healthcare	www.medical.philips.com
St. Jude Medical	www.sjm.com
Stryker	www.stryker.com
3M Healthcare	www.3m.com

Table 13-4. *Top Contract Research Organizations*

Accelovance	www.accelovance.com
Advanced Biomedical Research Inc.	www.abr-pharma.com
Averion International Corp.	www.averionintl.com
Cardiovascular Clinical Studies, Inc.	www.ccstrials.com
Chiltern International Inc.	www.chiltern.com
ClinDatrix, Inc.	www.clindatrix.com
Clinsys Clinical Research, Inc.	www.clinsyscro.com
Covance Inc.	www.covance.com
Criterium, Inc.	www.criteriuminc.com
ECRON-AcuNova	www.ecron.com
Encorium Group	www.encorium.com
Global Drug Safety & Surveillance Inc.	www.globaldss.net
Global Research Services	www.grs-cro.com
i3	www.i3global.com
ICON Clinical Research	www.iconclinical.com
INC Research	www.incresearch.com
Integrated Research Inc.	www.iricanada.com
Integrium, LLC	www.integrium.com
Kendle International	www.kendle.com

(*continued*)

Table 13-4. *Continued*

Medpace	www.medpace.com
MedTrials, Inc.	www.medtrials.com
MDS Pharma Services	www.mdsps.com
MSOURCE	www.msource-cro.com
Omnicare Clinical Research	www.omnicarecr.com
Paragon Biomedical, Inc.	www.parabio.com
Parexel International	www.parexel.com
PharmaNet Development Group	www.pharmanet.com
PPD	www.ppdi.com
PRA International	www.prainternational.com
Promedica International	www.promedica-intl.com
ProTrials Research, Inc.	www.protrials.com
Quintiles Transnational Corp	www.quintiles.com
ReSearch Pharmaceutical Services, Inc.	www.rpsweb.com
Synteract, Inc.	www.synteract.com

Professional organizations and a variety of educational groups offer online clinical training courses, and some award specialist certificates or continuing education credits (see Tables 13-6 and 13-7). Professional organizations are a good way to network with people who are already established in a clinical specialty field, and some offer student memberships. Training courses are a good way to supplement your knowledge of good clinical practices, clinical study operations, drug and medical device development procedures, medical terminology, and regulatory requirements. If you are interested in a specific clinical subspecialty, you may also

Table 13-5. *Temporary Workforce Agencies and Websites Advertizing in Clinical Development Positions*

ACT-1	www.act-1.com
BioSpace	www.biospace.com
Greylock Recruiting	www.greylock-recruiting.com
i3 Pharma Resourcing	www.i3pharmaresourcing.com
KFORCE Inc.	www.kforce.com
PharmaFORCE	www.pharmaforce.net
Medzilla	www.medzilla.com
PharmiWeb	www.pharmiweb.com
Science Careers	http://sciencecareers.sciencemag.org
Science Careers Web	www.sciencecareersweb.net

Table 13-6. *Professional and Trade Organizations*

Pharmaceutical	Pharmaceutical Research and Manufacturers of America (PhRMA)	www.phrma.org
	Drug Information Association (DIA)	www.diahome.org
	American Pharmacists Association (APhA)	www.aphanet.org
	CenterWatch	www.centerwatch.com
Biotechnology	Biotechnology Industry Organization (BIO)	www.bio.org
Medical Devices	Association for the Advancement of Medical Instrumentation (AAMI)	www.aami.org
	Medical Device & Diagnostic Industry (MD&DI)	www.devicelink.com
Clinical Research Organizations	Association of Clinical Research Organizations (ACRO)	www.acrohealth.org
Regulatory	Regulatory Affairs Professional Society (RAPS)	www.raps.org
	Food and Drug Law Institute (FDLI)	www.fdli.org
Project Management	Project Management Institute (PMI)	www.pmi.org
Clinical Research Professionals	Society of Clinical Research Associates (SoCRA)	www.socra.org
	Association of Clinical Research Professionals (ACRP)	www.acrpnet.org
Data Management	Society for Clinical Data Management (SCDM)	www.scdm.org
Biostatistics	American Statistical Association (ASA)	www.amstat.org
	Statisticians in the Pharmaceutical Industry (PSI)	www.psiweb.org
	International Society for Pharmacoepidemiology (ISPE)	www.pharmacoepi.org
	Biometrics Society	www.tibs.org
Clinical Quality Assurance	American Society of Quality (ASQ)	www.asq.org
Medical Writing	American Medical Writers Association (AMWA)	www.amwa.org
	International Society of Medical Publication Professionals (ISMPP)	www.ismpp.org
	Society for Technical Communication (STC)	www.stc.org
Consumer Healthcare Products	Consumer Healthcare Products Association (CHPA)	www.chpa-info.org

Table 13-7. Online Standards, Training, and Certification Programs

Regulatory and Compliance Standards	International Conference on Harmonization (ICH)	www.ich.org
	International Organization of Standardization (ISO)	www.iso.org
	Food and Drug Administration (FDA)	www.fda.gov
	European Medicines Agency (EMEA)	www.ema.europa.eu
	European Commission/medical devices	http://ec.europa.eu/enterprise/sectors/medical-devices/index_en.htm
	Compliance Online	www.complianceonline.com
	ClinfoSource	www.clinfosource.com
	Kriger Research Center International	www.krigerinternational.com
Regulatory Affairs	Regulatory Affairs Professional Society (RAPS) (regulatory affairs certification)	www.raps.org
	Drug Information Association (DIA)	www.diahome.org
	Medical Design and Manufacturing (MD&D)	www.devicelink.com
Project Management	Project Management Institute (PMI) (project management professional certification)	www.pmi.org
	Drug Information Association (DIA) (project management certificate)	www.diahome.org
Clinical Research Associate	Association of Clinical Research Professionals (ACRP) (CRA certification)	www.acrpnet.org
	Society of Clinical Research Associates (SoCRA) (certified clinical research professional)	www.socra.org
	Drug Information Association (DIA)	www.diahome.org
	ClinfoSource (CRA certificate program)	www.clinfosource.com
	Kriger Research Center International (CRA certificate program)	www.krigerinternational.com

(continued)

Table 13-7. Continued

Clinical Coding	Medical Dictionary for Regulatory Activities (MedDRA) (certifed MedDRA coder program)	www.meddramsso.com
	Clinical Data Interchange Standards Consortium (CDISC)	www.cdisc.org
Study Coordinator	Association of Clinical Research Professionals (ACRP) (clinical research coordinator certification)	www.acrpnet.org
	ClinfoSource (study coordinator certification)	www.clinfosource.com
	Kriger Research Center International	www.krigerinternational.com
Drug Development Overview	Drug Information Association (DIA)	www.diahome.org
	ClinfoSource	www.clinfosource.com
	Regulatory Affairs Professional Society (RAPS)	www.raps.org
Phlebotomy	Phlebotomy Central	www.phlebotomy.com
Medical Devices	Association for the Advancement of Medical Instrumentation (AAMI) (international certification commission certification)	www.aami.org
	Regulatory Affairs Professional Society (RAPS) (RAC medical device certification)	www.raps.org
	Medical Device & Diagnostic Industry (MD&DI) (white papers on various topics)	www.devicelink.com
Clinical Data Management	Society for Clinical Data Management (SCDM) (certified clinical data manager)	www.scdm.org
	Kriger Research Center International (data management certificate program)	www.krigerinternational.com
Statistical Programming	SAS (SAS certified professional)	www.sas.com
	Kriger Research Center International	www.krigerinternational.com
Quality Assurance	Society for Quality Assurance (SQA) (registry of quality assurance professionals)	www.sqa.org
	Kriger Research Center International	www.krigerinternational.com

Category	Resource	Website
Document Management Systems	Documentum (EMC Proven Professional certification)	www.documentum.com
Good Clinical Practice	ClinfoSource	www.clinfosource.com
	Society of Clinical Research Associates (SoCRA)	www.socra.org
	Association of Clinical Research Professionals (ACRP)	www.acrpnet.org
	Regulatory Affairs Professional Society (RAPS)	www.raps.org
	Society for Quality Assurance (SQA)	www.sqa.org
	Kriger Research Center International	www.krigerinternational.com
	Institute of Clinical Research	www.instituteofclinicalresearch.com
	Online GCP Training	www.gcptraining.org.uk
	Compliance Online	www.complianceonline.com
Medical Dictionaries and Terminology	ClinfoSource	www.clinfosource.com
	Kriger Research Center International	www.krigerinternational.com
	Free Ed Net	www.free-ed.net/free-ed/healthcare/medterm-v02.asp
	Des Moines University	www.dmu.edu/medterms
Clinical Safety	Drug Information Association (DIA) (clinical safety & pharmacovigilance certificate)	www.diahome.org
Clinical Study Operations	ClinfoSource	www.clinfosource.com
	Society of Clinical Research Associates (SoCRA)	www.socra.org
	Association of Clinical Research Professionals (ACRP)	www.acrpnet.org
	Drug Information Association (DIA) (clinical research certificate)	www.diahome.org
	Regulatory Affairs Professional Society (RAPS)	www.raps.org
Standard Operating Procedures	Society of Clinical Research Associates (SoCRA)	www.socra.org
	ClinfoSource	www.clinfosource.com

Table 13-8. *Salary Surveys*

American Association of Pharmaceutical Sciences Salary Survey	www.aaps.org
American Medical Writers Association Salary Survey	www.amwa.org
American Society for Quality Salary Survey	www.asq.org
American Statistical Association	www.amstat.org
Applied Clinical Trials Salary Survey	www.appliedclinicaltrialsonline.com
Association of Clinical Research Professionals CRA Salary Survey	www.acrpnet.org
Contract Pharma Salary Survey	www.contractpharma.com
Medical Device & Diagnostic Industry Salary Survey	www.devicelink.com
Project Management Institute Salary Survey	www.pmi.org
Salary.com	www.salary.com
Society for Technical Communication Salary Database	www.stc.org

want to pursue online training for skills such as SAS programming, Medical Dictionary for Regulatory Activities (MedDRA) coding, electronic document management systems, or quality assurance auditing. For further information about the courses and certifications offered by these organizations, consult the Resources section of Chapters 3–10.

Glossary

Adverse event (AE) Any unwanted reaction, side effect, or other event experienced by a patient during the course of a clinical study, which may or may not be associated with use of the experimental product.

Annotated case report form A blank case report form (CRF) that includes instructions for the format, units of measure, and types of information to be entered in each field of the CRF.

Audit A systematic and independent examination of study-related activities and documents to determine whether the evaluated study activities were conducted, and the data recorded, analyzed, and accurately reported, according to the protocol, sponsor's standard operating procedures (SOPs), and the applicable regulatory requirements.

Audit report A written evaluation by an auditor communicating the results of an audit.

Audit trail Documentation about who, when, and what data changes were made to a clinical database, allowing auditors to reconstruct the course of events during data entry and query resolution.

Bias A situation in which the collection, analysis, interpretation, or review of data leads to conclusions that are prejudiced.

Biopharma A collective term referring to both biotechnology and pharmaceutical drug development companies.

Biotechnology A collection of technologies that capitalize on the attributes of cells, such as their manufacturing capabilities, and make therapeutic products from biological molecules, such as DNA and proteins.

Blinding The process of concealing treatment assignments to avoid potential bias in the recording and evaluation of results. Single blinding usually means the patients are unaware, and double blinding usually means the patients, investigators, monitor, and in some cases data analysts are unaware of the treatment assignment(s).

Case report form (CRF) A printed, optical, or electronic document designed to record all of the protocol-required data from each study patient. The forms are completed by staff at the investigator site and submitted to the sponsor.

Class III medical device Highly regulated medical devices used for life-support or life-sustaining purposes, such as heart valves and implantable pacemakers. Because they pose a potentially unreasonable risk of illness or injury, Class III medical devices must be evaluated in clinical studies to obtain market approval.

Clinical Pertaining to or founded on direct observation and treatment of patients.

Clinical Data Interchange Standards Consortium (CDISC) A global nonprofit organization that sets platform-independent data standards for acquisition, exchange, submission, and archiving clinical data.

Clinical quality assurance (CQA) The corporate department or function that uses auditing, quality assurance (QA) consultation, and other QA activities to assess and support compliance with good clinical practice and other regulatory standards.

Clinical site Location where a clinical study is conducted. Clinical sites may be academic institutions, hospitals, or private clinics. The clinical site facilities and staff must be suitable and qualified to carry out the study as specified in the clinical protocol.

Clinical study An investigation of patients to discover or verify the clinical, pharmacological, and/or therapeutic effects of an experimental medical product, and/or to identify any adverse reactions to an experimental medical product.

Clinical study report (CSR) A written document describing the clinical treatments and procedures, statistical analysis methods, key results, and conclusions of a clinical study.

Clinical trial *See clinical study*

Clinical Trial Authorization (CTA) A European application requesting permission to test a new product in patients.

Common Technical Document (CTD) A format for organizing applications to regulatory authorities for market approval of pharmaceuticals for human use. The requirements for CTDs were developed and are maintained by the International Conference on Harmonization (ICH).

Concomitant medication A drug given at the same time, or almost at the same time (such as one after the other or on the same day), as another drug.

Continuing medical education (CME) credits Supplemental education of healthcare professionals to maintain competence and learn about new and developing clinical areas. The activities may be live events, written publications, or programs using electronic media. Credits are earned at one or two credits per hour of learning activity, and many healthcare licensing boards specify CME credit requirements in order to maintain a license.

Contract research organization (CRO) An independent organization contracted by the sponsor to perform one or more of a sponsor's study-related duties and functions.

Data monitoring committee (DMC) An independent committee that may be established by the sponsor to assess the safety data, critical efficacy end points, and the progress of a clinical study. DMCs are empowered to recommend continuing, modifying, or stopping the study.

Database lock An action that prevents further changes to a clinical study database. A database is locked after a determination has been made that query resolution is complete and the database is ready for analysis.

Double blind *See blinding*

Double data entry A quality control method in which a second person enters the same data as the original data entry worker and the two entries are compared to verify they are the same. If the two entries are different, the operators follow a standard procedure for resolving the discrepancy. Only double-verified (or resolved) data pass into the database.

Edit checks An auditable process for checking the data in the clinical database, either electronically or manually, to ensure that each data value is complete, within a reasonable range, medically plausible, consistent with other data and patient information, and complies with the protocol.

Efficacy A beneficial effect of a drug or treatment, producing either an improvement in the course or duration of a disease.

Electronic data capture (EDC) A computerized system designed to collect data from a clinical study in an electronic format.

End point A symptom, sign, or laboratory abnormality that constitutes the means by which the clinical study objectives are measured. For example, a common end point in cancer studies is disease progression.

Essential documents Documents that individually and collectively permit evaluation of the conduct of a study and the quality of the data produced. These documents serve to show that the investigator, sponsor, and monitor have complied with Good Clinical Practices and all applicable regulatory requirements.

Ethics committee *See institutional review board*

EudraCT number A unique number used to identify a clinical study that is conducted in one or more countries in the European Community.

European Medicines Agency (EMEA) A decentralized body of the European Union with headquarters in London. Its main responsibility is the protection and promotion of public and animal health, through the evaluation and supervision of medicines for human and veterinary use.

Exclusion criteria The characteristics that would prevent a patient from participating in a clinical study, as outlined in the clinical protocol.

Food and Drug Administration (FDA) An agency of the United States Department of Health and Human Services responsible for safety regulation of foods, dietary supplements, drugs, vaccines, biological medical products, blood products, medical devices, radiation-emitting devices, veterinary products, and cosmetics.

Good Clinical Practice (GCP) An international ethical and scientific quality standard for the design, conduct, performance, monitoring, auditing, recording, analyses, and reporting of clinical studies. GCPs provide assurance that the data and reported results are credible and accurate and that patients' rights, integrity, and confidentiality are protected.

Inclusion criteria The characteristics that must be met by a patient to participate in a clinical study, as outlined in the clinical protocol.

Independent ethics committee *See institutional review board*

Informed consent A process by which a patient voluntarily confirms his or her willingness to participate in a clinical study, after having been informed of all aspects of the study that are relevant to the patient's decision to participate.

Informed consent form (ICF) A document that provides information, in lay terms, to the patient about the experimental product and the clinical study so that the patient can assess the anticipated risks and benefits of participating in the study. The signed and dated document signifies the patient's informed consent to participate in a clinical study.

Institutional review board (IRB) An independent body, constituted of medical/scientific professionals and nonmedical/nonscientific members, whose responsibility is to ensure the protection of the rights, safety, and well-being of patients involved in a clinical study and to provide public assurance of that protection, by reviewing and approving the study protocol, the suitability of the principal investigators (PIs), the adequacy of the facilities, and the methods for obtaining and documenting informed consent of the study patients.

Interactive voice response system (IVRS) An interactive technology that allows a computer to detect voice and keypad inputs and respond with prerecorded or dynamically generated messages to direct users on how to proceed.

Interim analysis Analysis of data generated in a clinical study prior to conducting a final analysis. A formal interim analysis is planned, explained in the statistical analysis section of the clinical protocol, and conducted while the study is in progress for the purpose of making a decision to continue or terminate the study.

International Conference on Harmonization (ICH) A project with regulatory and pharmaceutical industry representatives from Europe, Japan, and the United States to harmonize technical guidelines and requirements for registration of human pharmaceuticals.

International Organization for Standardization (ISO) An international standard-setting body recognized by many governments for industrial and commercial standards.

Investigational Device Exemption (IDE) Regulatory documents that are submitted to and approved by the Food and Drug Administration (FDA) before starting clinical studies using an experimental medical device in the United States.

Investigational New Drug (IND) application An application containing a sponsor's plans and accumulated laboratory data and petitioning the FDA to allow clinical studies using an experimental drug in the United States.

Investigator site *See clinical site*

Investigator's brochure (IB) A compilation of the accumulated clinical and non-clinical data on an experimental product for reference by the principal investigator during a clinical study.

ISO 9000 A family of standards for quality management systems, which is maintained by the International Organization for Standardization (ISO). Companies that have been audited and certified can publicly state that they have been ISO certified or ISO registered, indicating that they meet ISO's standards for quality systems.

Lead clinical research associate (CRA) A senior-level clinical research associate responsible for coordinating the activities of a group of CRAs, who are typically needed to monitor the clinical sites of large clinical studies. The lead CRA may monitor a few sites but is primarily office-based and serves as a liaison between the clinical study team and the CRA monitors.

Marketing Authorization Application (MAA) An application, in a common technical document format, that must be reviewed and approved by a regulatory agency before the medical product can be marketed in that country.

Medical device Any instrument, apparatus, software, or other material intended for use in people for diagnosis, prevention, monitoring, or treatment of a disease, injury, or handicap.

Monitoring The act of overseeing the progress of a clinical study and of ensuring that the study is conducted, the data are recorded, and the study status is reported in accordance with the protocol, standard operating procedures (SOPs), Good Clinical Practice (GCP), and the applicable regulatory requirements.

New Drug Application (NDA) An application to the FDA for a license to market a new drug in the United States. The NDA must include data from all studies conducted using the experimental drug, demonstrating its efficacy and safety. *See also marketing authorization application*

Patient listings Tables of data drawn from the clinical study database and listing individual data points from all patients in the study. Patient listings in a clinical study report include tables showing the discontinued patients, protocol deviations, patients excluded from the efficacy analysis, individual efficacy response data, adverse events of each patient, and individual laboratory measurements by patient.

Patient narratives A brief description of a serious adverse event or other observations of special interest because of their clinical importance. The description includes the nature and intensity of the event, the clinical course leading up to the event, the timing of experimental treatment, relevant laboratory measurements, whether treatment was stopped and when, and measures taken to protect patient safety. The narrative also includes a medical opinion on whether the event was caused by the experimental treatment. Patient narratives are written by a physician or safety specialist and are included in regulatory safety reports and clinical study reports.

Placebo A pharmaceutical preparation that does not contain the experimental drug. In blinded studies, placebos are prepared to be physically indistinguishable from the preparation containing the experimental drug.

Premarket Approval Application (PMA) An application submitted to the FDA for evaluation of safety and effectiveness and market approval of Class III medical devices.

Principal investigator (PI) An appropriately qualified person (usually a physician) responsible for conducting a clinical study at a clinical site. If the site includes other investigators who assist with the study, the PI retains overall responsibility and accountability for following the protocol and treating the patients.

Product dossier An evidence-based evaluation that summarizes the properties, claims, indications, and conditions of use for a drug in a factual, straightforward manner. Product dossiers are distributed to health systems or managed-care organizations for their reference.

Protocol A document that describes the objectives, design, methods, statistical considerations, and organization of the study. Clinical protocols usually include the background and rationale for the study.

Protocol synopsis A summary of the clinical protocol used in preliminary discussions with potential investigator sites. The protocol synopsis typically includes background information about the experimental treatment, the study's objectives and design, patient selection criteria, study procedures, and key data to be collected.

Quality assurance (QA) The systematic and independent examination to assure that all study-related data and documents sent to regulatory agencies are true and factual. QA audits determine whether the evaluated activities were appropriately conducted and data were collected, documented, analyzed, and reported in compliance with the protocol, GCPs, and the sponsor's SOPs.

Quality control (QC) Checking operational techniques and activities at appropriate intervals to verify that the output of the clinical study (such as the data and clinical study report) meet the requirements stated in the protocol, GCPs, and the sponsor's SOPs.

Query A request for clarification on a data item that is discovered during data review to be incomplete, inconsistent, or illogical.

Query resolution Final clarification by the PI of a query posed by the sponsor or sponsor's representative.

Randomization The process of assigning study patients to treatment or control groups using an element of chance so that the assignments reduce treatment bias.

Regulations Rules prepared by government agencies and used to administer a law.

Regulatory agencies Government bodies having the power to regulate, including the review of submitted clinical data and market approval of new medical products.

Safety An assessment of the medical risk to the patient from use of a medical product, usually determined in a clinical study by laboratory test results, vital signs measurements, and reports of clinical adverse events.

SAS An integrated system of software products provided by SAS Institute that enables a programmer to perform data management, statistical analysis, data warehousing, and other manipulations of clinical data.

Serious adverse event (SAE) Any adverse event that results in death, is life-threatening, requires hospitalization, or results in persistent or significant disability of the patient.

Single blind *See blinding*

Six Sigma A business management strategy that seeks to identify and remove the causes of defects and errors in business practices using a defined sequence of steps and quantifiable financial targets.

Source document A document in which data collected for a clinical study are first recorded, generally the patient's medical record, laboratory instrument printouts, and other original records such as X-ray images.

Sponsor An individual, company, institution, or organization that takes responsibility for initiation, management, and/or financing of a clinical study.

Standard operating procedure (SOP) Written, step-by-step instructions to achieve uniformity in the performance of a specific function.

Statistical analysis plan (SAP) A document that contains a detailed, technical description of the statistical procedures, consistent with the objectives of the clinical protocol, to be used in analyzing the data from a clinical study for evidence of a medical product's efficacy, safety, and other properties.

Stopping rules Defined safety criteria that, if met, will either pause or halt the clinical study, because of futility of the study or risks to the study participants.

Structure Query Language (SQL) An interactive programming language for querying and modifying data and managing databases. SQL can be used to create databases and retrieve and manage data in relational database systems.

Study coordinator An appropriately qualified person who handles most of the administrative responsibilities of a clinical study on behalf of the PI, serves as a routine point of contact at the site, and maintains all data and records for review by study monitors.

Synopsis A brief overview prepared at the conclusion of a clinical study, summarizing the study plan, objectives, inclusion criteria, methodology, and results. *See also protocol synopsis*

Total Quality Management (TQM) A management approach centered on quality, based on participation of all its members, and aiming at long-term success through customer satisfaction.

Unanticipated adverse device effect (UADE) A serious and unanticipated adverse event associated with a medical device or any other unanticipated serious problem associated with the device that affects the patient's safety or welfare.

User acceptance testing (UAT) Verification by users of the proper functioning of a clinical database or EDC system using test data that simulate the types of data that will be collected in the study. UAT is one of the final stages of database and EDC development before the software is released for use in a clinical study.

Vital signs Various physiological measurements such as body temperature, pulse, blood pressure, and respiratory rate, to assess basic body functions.

Index

About the Author

Rebecca J. Anderson has worked in the bio-pharmaceutical industry for more than 25 years in jobs spanning pharmaceuticals, biotechnology, medical devices, and contract research organizations. As a hiring manager in all of these organizations, she has reviewed and screened thousands of applicants for clinical research positions. Dr. Anderson holds a Ph.D. in pharmacology from Georgetown University and was

Courtesy of David Roth Photography.

an MRC Postdoctoral Fellow at the University of Toronto. Prior to her career in industry, she conducted basic research in pharmacology and toxicology, and served on the faculties of the George Washington University Medical Center and the University of Michigan School of Public Health. She currently works as a freelance writer.